THE FAMILY GENE

THE
FAMILY
GENE

*A Mission to Turn My Deadly Inheritance
into a Hopeful Future*

JOSELIN LINDER

An Imprint of *HarperCollins*Publishers

HarperCollins books may be purchased for educational, business, or sales promotional use. For information, please e-mail the Special Markets Department at SPsales@ harpercollins.com.

FIRST EDITION

Designed by Jane Treuhaft

Library of Congress Cataloging-in-Publication Data has been applied for.

ISBN 978-0-06-237889-7

17 18 19 20 21   RRD   10 9 8 7 6 5 4 3 2 1

"HARVEST" (LYRICS)

What can you say
About one who has gone away?
You may curse
You may frown
You may ask
Hey man—why don't you come on back?

What can one do
About he who has left you?
One can cry
One can pray
One can ask
Hey man—why don't you come on back?

Dream on I am told
The cards have been folded
The movie's over
The ship has passed
The day is gone
It's night at last

You can relive what has
Already passed
Don't you feel it?
Have you seen
Though he has gone
He will always have been

BILL LINDER
1947–1996

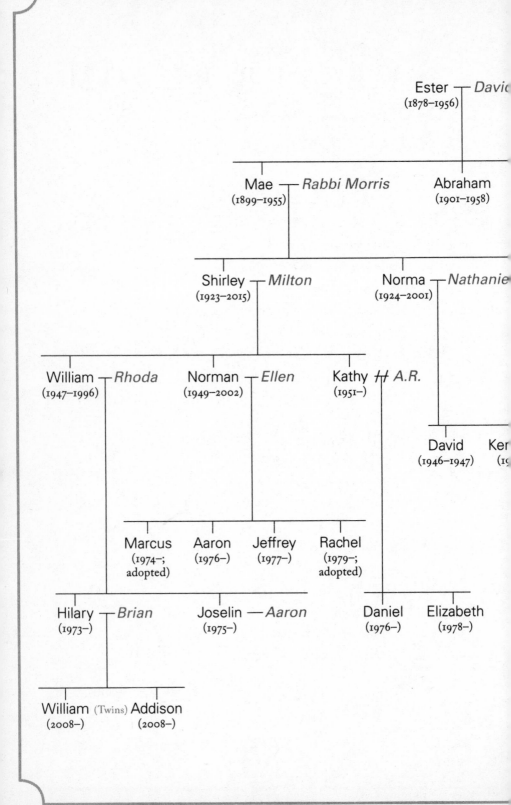

DESCENDANTS
of
ESTER BLOOM

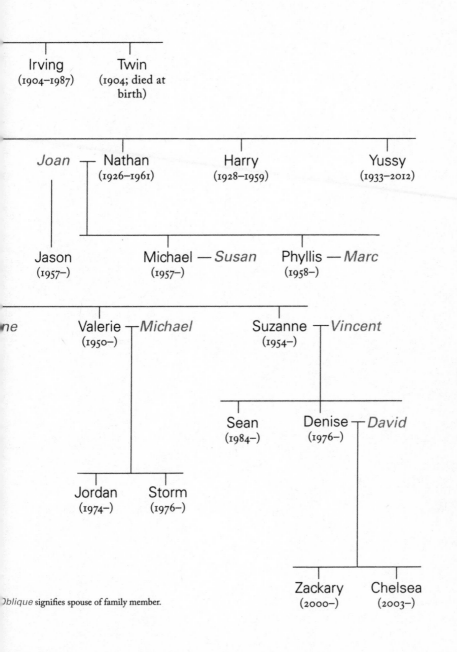

Irving
(1904–1987)

Twin
(1904; died at
birth)

Joan — Nathan
(1926–1961)

Harry
(1928–1959)

Yussy
(1933–2012)

Jason
(1957–)

Michael — *Susan*
(1957–)

Phyllis — *Marc*
(1958–)

ne

Valerie — *Michael*
(1950–)

Suzanne — *Vincent*
(1954–)

Sean
(1984–)

Denise — *David*
(1976–)

Jordan
(1974–)

Storm
(1976–)

Zackary
(2000–)

Chelsea
(2003–)

Oblique signifies spouse of family member.

INTRODUCTION

When I was sixteen years old, my dad sat me down in the living room and told me he was very sick. My mom was away and my sister had recently left our home in Columbus, Ohio, for college in Connecticut.

"What do you have?" I asked.

"I don't know," he answered. He looked terrified.

"Then how do you know it's bad?" I asked sincerely. My dad had been complaining of fatigue. I remembered him fainting the summer after a family vacation, but he chalked it up to "getting up too fast." Otherwise, he was a tall, healthy-seeming, forty-four-year-old man. He didn't look pale. He wasn't running a fever.

"I want your mother," he said.

I didn't know what else I could do, so I called my mother and told her to come home.

———

Five years later, on September 24, 1996, my father succumbed to his illness. After he died, the surgeon who assisted with the autopsy described his organs as "practically fused together." They had been pressed so tightly against one another by countless liters of fatty lymph fluid that they had basically congealed. Doctors had only ever encountered this illness twice before—in my dad's fifty-five-year-old grandmother (who'd passed away in 1955) and in his thirty-four-year-old uncle (who followed her in 1961). No one drew a connection between the three illnesses until two years into my father's decline.

In the years that followed, several of my family members experienced similarly strange symptoms. After tests revealed a complex system of swollen veins throughout her digestive tract, my thirty-something sister was told by her doctors that she was "a ticking time bomb." The same year, my father's first cousin lay in a coma after a vein burst in her brain. It took three years before she could walk and talk again. At a routine checkup, another cousin was told that she needed a liver transplant as soon as possible.

I wasn't spared. When I was in my twenties, my legs started to swell. Every time I went to the doctor, I received ominous news—from reports of extreme liver blockage to increased quantities of wayward fluid much like my father's, deposited where it should not be deposited within my body. In my thirties, I was plagued with questions about my reproductive capabilities. Not a single medical professional could explain what was causing such terrible complications in an otherwise healthy family.

My father's illness was unique in the medical field, and it attracted the attention of a Harvard geneticist who has spent the last twenty-one years trying to find out what was killing him, and if it was killing us. Together, we discovered that my father's body was reacting to the expression of a genetic mutation, now more commonly called a genetic *variant*, that seemed to have occurred only a few generations earlier—a private mutation. Private mutations are genetic variants found in a group of related individuals. For one of the first times in human medical history, my family is allowing medical science to witness a "founder effect"—a new mutation passing to subsequent generations during its fifth generation, instead of hundreds or even thousands of generations after the initial mutation of the gene.

Medical genetics has given my family a way of addressing our illness. We now understand the gene and its impact on our bodies. For us, this means hope, and the chance to change our fate. Medical mysteries like these are rarely solved. The fact that we've made so much progress in seventeen years has a profound significance that is not lost on any of us. My family has an opportunity to weed a deadly gene out of existence, to use the power of medicine to do something that humanity has never before been presented with the opportunity to do. But nothing is certain.

Our story is the story of science, its shortcomings, and its miraculous capabilities to change the world. It's the story of doctors on a quest to map a gene and understand it. It's the story of our gene—the lives it claimed, the lives it has changed, and the future of genetics.

ONE

It began the summer after I turned fifteen in 1990, just after our parents took Hilary and me on a two-week trip to Israel. Our great-aunt, who had joined us on the trip, suffered a minor heart attack. Our mother stayed behind to care for her while Hilary and I flew home to Columbus with our dad.

While my father was napping away his jet lag upstairs, Hilary and I set ourselves up in the basement to watch *Dirty Dancing* for the millionth time on our new VCR. As Johnny lifted a dripping white-tank-top-wearing Baby powerfully over his head in the lake, a loud series of thuds echoed down the basement stairs.

Hilary jumped up first. She was a year and a half older than me and nearly seventeen. She took the stairs two at a time out of the basement, her long athletic legs confident as she raced around the corner to a second set of stairs up to the second floor. I tiptoed behind. Hilary landed next to our father's head at the top of the stairway. His eyes blinked, glazed with confusion. Blood dripped from above his forehead to his cheek, forming a rivulet through his dark hair.

"Come here and stay with him," my sister ordered.

"Where are you going?" I called up from where I stood at the bottom of the stairs.

"I'm calling an ambulance," she answered. "Get up here."

My father was on his stomach at the top of the stairway with his head near-tangled in the metal bars of the banister. That banister had kept him from falling headfirst onto the stone-tile floor below, but also gashed open his scalp. I approached him carefully, winding my way up around the curve in the steps.

"Get him a towel," Hilary called while explaining the situation to 911. "Daddy! Are you okay?" she shouted.

He was sitting up now. I fumbled a towel off the rack in the nearby bathroom I shared with my sister.

"Yes," he answered quietly. I watched awareness settle over him. He suddenly focused on my face, no longer gazing out vaguely at nothing in the distance. Slowly he knitted his brow in an expression of absolute confusion, or was it disgust? At first I wasn't sure what was going through his head. For all I knew, he was having a stroke.

Then it dawned on me. I was hysterical. My now-sober father was trying to assess not just his own medical state, but mine too. I hadn't noticed earlier since all I could hear was the thumping silence of residual panic born solely from the possibility that my father had been in mortal danger.

My dad rolled his eyes and tossed the towel I'd given him back at me, presumably so I could wipe the tears that were pouring down my own face. Hilary returned and helped our father to his feet. He was composed enough to put himself back to bed, order us to cancel the ambulance, and request a glass of water. Meanwhile, I went to the bathroom and took several deep breaths.

An alarm had been sounded, but right then it was unclear what it foretold—or that it even had anything to foretell. We were not a family who routinely dealt with catastrophe. We lived in Ohio.

My dad always liked watching thunderstorms. Once we had all

been sitting on our front porch watching a particularly wild storm when lightning struck our neighbors' house across the street. I remember the electricity tickling the tiny hairs on my face as sparks flew off their roof. We went inside and called them in time for a fire truck to come and contain the fire to the attic. We were heroes that day, but really it's just what we were good at. It's where we excelled: watching lightning strike other people's houses.

We called our dad's nearby office, where he worked as a physician. Hilary drove when we accompanied him to get seven stitches in his head.

As the nurse cleaned the wound, he explained that he had gotten up too fast from his nap. He explained that when we arrived home from the airport, he had taken a diuretic, or a water pill, designed to help reduce swelling from water weight. He was jet-lagged and otherwise tired from our two-week trip. As he stood up too quickly from his nap, he became disoriented and then finally fell at the top of the stairs.

My dad left out several important items that didn't come to light until much later: two months before our trip to Israel, in April of 1990, he went to see Dr. Keith Pattison, a cardiologist, because of minor swelling in his ankles. Ankle edema, as it is known, is often linked to heart disease. My father had undergone open-heart surgery as an adolescent in the 1950s. Dr. Pattison offered no definite medical reasons for the swelling that day. His findings simply maintained that "the patient is not functionally impaired . . . and he retains superb exercise capacity." In other words, my dad's ankles might be swollen, but they didn't seem to affect his heart or health.

Dr. Pattison more or less chalked up those puffy ankles to a familial trait. My dad's mother and maternal aunt both experienced similar swelling and they were healthy. To my father's mind, that was probably explanation enough. He seemed to have felt pretty good otherwise. He'd videotaped our whole trip and danced to klezmer music at least once.

Pressures on board the aircraft during the flight home from Israel likely impacted the pressures in his body, causing his ankle swelling to expand upward into his legs. Those cabin pressures are why plastic water bottles crinkle up in flight.

At the time I don't recall knowing about that doctor's visit or the swelling in my dad's legs at least in part because I was a teenager. It certainly wasn't of any concern in my day-to-day life. Instead, I went with the story my dad, his head wound now stitched, told us that day. He'd taken a water pill for swelling he'd gotten on the plane, which, at the time, sounded like something people sometimes got on planes, like summer colds and tomato juice. My dad seemed healthy. He was active and even athletic. From my perspective, after the stitches had been stitched, it truly seemed the worst was over. There was absolutely no reason to think otherwise.

TWO

Two years later, during my senior year of high school, when I was arguably at the zenith of my adolescent narcissism, my father's health concerns—if there were any—existed only at the far corners of familial conversation. He sometimes mentioned feeling tired and had grown fascinated with a trendy illness called chronic fatigue syndrome. But he seemed okay. After all, he'd recently started a band that met in our basement that was appropriately embarrassing to high schoolers, and he'd purchased a rowing machine.

When a beloved ice cream chain from Cincinnati called Graeter's opened up near our house in Columbus, my father wanted us all to go there for our first time together during Hilary's semester break at the University of Hartford, where she was a sophomore. My parents' best friends, Jackie and Jimmy, joined us, as they often did for family outings.

When we arrived at the counter we called out our orders one at a time. When the ice cream server got to our mother, his expression changed.

"Oh, hey, Rhoda!" he said informally.

"Hey," our mother muttered under her breath.

"Can I get you your usual?"

We were all stunned. To be fair, Graeter's is the undisputed producer of the world's greatest mint-chocolate-chip ice cream. But the place had literally *just* opened: it was at most a month old. How could this person know my mother's name, much less her *usual*? My mom's love of sweets wasn't a secret, but even for her, this was fast work. She sheepishly admitted she had been going to Graeter's every day, sometimes twice a day, since it had opened. And although that day we all laughed about it, the incident remained troubling to my father. My mother wasn't overweight, but she agreed she couldn't always control her intense sugar cravings.

My dad was pathologically weight conscious. The product of a family who prized a healthy, skinny body, he immediately began asking my mother to go *for treatment*. She later told me that this pressure became a point of real contention in their marriage. It's possible my dad was fat-phobic, not that he had a problem with overweight people, per se. But over the years, more than one close female friend of his admitted to me, unprompted, that he was always fairly casual about mentioning it when he thought they were putting on weight. While it might have been related to my dad's lifelong commitment to medicine and health, some part of it was certainly linked to his upbringing and a subsequent mind-set that mirrored his parents' that people, especially women, should be thin. After a year of fighting, my mom finally agreed to sign herself up for a reputable Texas wellness spa—that we had gleefully taken to calling "fat camp"—to fight any future weight gain.

The day she left, two years since my father's fall at the top of the stairs after our family trip to Israel, she couldn't have known that my father's body was ready to deliver its next harsh blow.

I don't remember when I first noticed my father's swelling. I once watched as he pulled off a pair of black stockings after a long day at the clinic. His lower legs appeared chickenlike beneath a hovering swollen knee. I had asked him about it and he told me it was normal for his legs to swell after a long day at work. He said the stockings he wore used pressure to help push the fluid toward his heart. I don't remember delving any more deeply into it than that.

Twenty years later, when I began researching the early days of my father's illness for a National Public Radio program that never aired, one of the first people I called was Kim Kneuven. Kimmy, as I always called her, had worked alongside my dad as a nurse since his first job out of medical school. Kim's shiny straight black hair and fair, peachy complexion went well with her warm laugh and Ohio twang.

Kim and my father were fast friends from the day they met, bonding over a heart procedure they had both undergone as adolescents that left them with long matching scars running down their chests. They were almost five years apart in age to the day, often sharing an office birthday party. Kim said that my dad always tried to get out of the standard doctor-nurse rotation so that he could continue to work with her. One day he made the suggestion that every doctor at the clinic pair off with one specific nurse. When he had everyone on board, since it was his idea, he chose first. He chose Kim.

Kim told me she'd noticed in her career that a lot of doctors didn't act like healers the way my father did. He was the kind of doctor who gave his patients our home phone number and never minded that they used it. As a doctor, he was hardworking and honest, and it helped that he was the kind of person who could get away with calling strangers "buddy." More than one of his patients' children was named after him.

When my dad left his first job to start up his own practice, Kim naturally went with him. Fifteen years into growing it, the clinic was thriving and my father had just opened up second and third locations.

On Fridays, Kim said they often stayed after the clinic closed to talk about the business over a beer. On one of these Fridays, Kim found my father sitting alone in an exam room. His breathing was shallow.

"I can't catch my breath," he told her. "I need you to take me to the ER."

"Shit, let's go!" Kim replied. With my mother away, Kim called my parents' best friends, Jackie and Jimmy, to meet them at the ER. Then she helped my dad, struggling to breathe, to her car.

Jackie left Jimmy at their house. The call from Kim sounded like my dad simply needed a lift back to his car. Jackie and Jimmy lived down the street from us and my father's office was only five minutes away. While Jackie was being seated in the ER waiting room, my dad was already being seen.

During his X-ray, the doctor saw something alarming. My father had what medical professionals call "pleural effusions": fluid in his lungs. Kim stood beside him during a procedure in which doctors removed the fluid by sucking it out through a tube placed into an incision leading directly into the lungs. The process is called "tapping"—as in maple syrup from a tree—and it's about as crude. What's worse, the patient is kept awake and conscious the entire time. Kim held my dad's hand. Both were terrified.

The doctor extracted almost a liter of a fluid the color of lemon meringue pie. Kim had never seen anything like it, and she'd been a nurse for more than twenty years.

"What the hell is that?" she asked.

My dad and the attending were both at a total loss.

Kim said all she could think was that he'd caught something on our Israel trip even though more than a year had already passed, perhaps because to Kim it seemed like his health had been on the decline ever since.

We later learned that what came out of his lungs that day was "lymphocytic exudate," or "chylous fluid"—a sticky, milky substance made up of lymph, emulsified fats, protein, and sometimes cells. It is also referred to by its perky nickname "chyle." Lymphatic fluid of any type should not ever leak into the lungs. It does, from time to time, but almost exclusively in extreme cases of heart disease, liver failure, and certain types of cancer (most typically ones that *cause* heart disease or liver failure). Consistency-wise, the fluid is saucelike, and when it's in the lungs, it can make it very difficult to breathe. Most importantly, the presence of chylous fluid outside the lymphatic system of a healthy forty-five-year-old man is almost completely impossible to account for.

Jackie, who had sat alone in the waiting room, drove my father home that night and helped him to bed.

The morning after his harrowing visit to the ER, my dad called me into the living room. Sunlight streamed in through the sliding-glass doors that led to our backyard.

My father looked scared.

"I want your mother," he said. "I'm sick."

"Sick with what?" I asked.

"I don't know," he replied. And although it wasn't something I would readily recognize in my father, I thought I could see tears welling up.

"Well," I ventured, "how do you know you're sick?"

He didn't seem sick. He hadn't been acting sick. As far as I knew, he'd never even had a cold. He didn't answer me.

I have no idea where I had been that night. Maybe sitting with my friends at a Denny's with a bottomless hot chocolate and a pack of Marlboro Lights. My dad and I sat together on that bright sunny morning trying not to meet each other's eyes.

Soon I rose and went into the kitchen, where my mom had left the number for the fat camp on the counter. I called it and I told her to come home.

THREE

Out of the gate, it seemed obvious to my father and his medical team that his problems had something to do with his heart. He bounced from cardiologist to cardiologist, even seeking out doctors who might still be living who could recall his childhood surgery. A link seemed inevitable since the buildup of chylous fluid could be symptomatic of serious heart disease, and my father bore a scar from neck to navel. In those first months following the ER visit, he immediately delved into that childhood illness, looking deep into his past to see if he and his medical team could cobble together some kind of correlation.

My dad was born and raised in Pittsfield, Massachusetts. Situated in the Berkshire Mountains of western Massachusetts, it's a fifteen-minute drive up Route 7 from Stockbridge, where Norman Rockwell lived and died.

According to my grandmother, my dad was born at Pittsfield's only hospital five weeks early. There, she couldn't produce enough milk for her preemie. Pressure from the doctors made her anxious.

Another new mother on the ward offered to nurse her baby for her. Quickly my dad began to thrive.

Other than those first few stressful days, his early childhood was unremarkable. My dad was a playful, energetic boy. When his brother, Norman, was born, my dad was an active toddler. Two years later, Kathy came along.

Pittsfield had one high school, one library, a Friendly's restaurant, two gas stations, and a converted covered wagon serving popcorn to eager children on weekend strolls through Park Square. My parents were among those children, each growing up with their family on a street called Robbins Avenue, in houses catty-corner to one another. My mother, two years older than my father, recalls pushing him in his stroller down the block their houses shared.

My dad's family lived with his paternal grandparents on a small dairy farm just a few blocks from everyone in their family. My father's maternal grandparents lived across the street.

He grew to his full six feet four inches while still in high school. Athletic and strong, he would have been a shoo-in for the basketball team at Pittsfield High, but he had never been allowed to play. That's because when he was twelve years old, he had undergone one of the nation's very first open-heart surgeries.

When he was born, my dad had been diagnosed with a somewhat common congenital heart defect called "pulmonary valve stenosis." Pulmonary valve stenosis involves a narrowing in the valve that connects the right ventricle—one of the four chambers of the heart: right and left atria on top and right and left ventricles below—to the pulmonary artery that carries blood to the lungs to pick up a load of oxygen. This narrowing causes what my father had always described as a sticky flap. The sticky flap in turn leads to a heart murmur.

A heart murmur isn't one specific thing. In fact it's actually just a sound—the sound of the heart doing something extra or different. Instead of a heartbeat sounding like this—*buh-bum*—it might sound like *buh-ss-bum* or *bup-hh-bum*. It's an indistinct sound within a heartbeat. And it doesn't mean any one specific thing either. Typically heart murmurs indicate a problem with blood flow either through or in or out of the heart, which, as the pump that keeps our literal lifeblood moving, could be alarming. But a lot of heart murmurs become classified as "functional" or "innocent" in that they don't represent an immediate cause for alarm and might even remain harmless for a lifetime.

It's impossible to say whether or not my father's murmur would have ultimately proved functional, or if the sound his doctors heard indicated a truly troubling condition. It wasn't uncommon for functional murmurs to be treated aggressively, since the alternative might prove fatal.

My dad's murmur probably included an extra sound in the middle of his heartbeat; a whoosh of air indicating trouble. As a result, he was sidelined from sports, which, for an active, competitive child, was a real drag. Otherwise, his murmur impacted very little in his first twelve years. My dad never felt sick or unwell. He simply had to avoid play that could lead to overexertion, and then, frankly, only when someone was there to stop him.

But the doctors maintained that my father's overall health made him a great candidate for a brand-new procedure that could unstick his sticky valve. In 1959, when he was twelve years old, his parents drove him from their home to Boston's City Hospital for open-heart surgery.

In addition to other kinds of rapid technological growth linked to World War II, a severe need for medical advancement encouraged the creation of improved techniques for health care. Modern surgery stood to profit from such new developments as anesthesia, antibiotics, and transfusions. As war raged around him, a young army surgeon named Dr. Dwight Harken found himself tasked with figuring out a way to save the many young servicemen arriving in his infirmary with shrapnel lodged in their hearts. On animals, he began practicing a technique for cutting into beating hearts and working his fingers around the chambers in order to dig out trapped bullet casings and shell fragments. He worked steadily until his success rate was around 65 percent, then he tried it out on a man. With the success of that procedure, he learned that the human heart could be cut and manipulated *and continue to beat.*

Once the war was over, doctors began to wonder if Harken's technique might be applicable to other fatal or dangerous heart conditions not found on battlefields, especially congenital heart defects in children. In 1948, Harken was one of the first surgeons to report a procedure in which he corrected a condition called "mitral stenosis," a narrowing of the mitral valve that passes blood between chambers of the heart—like a door between your living room and the kitchen that just doesn't open wide enough to fit your hungry guests through it at Thanksgiving. Using an identical technique to the one he had used to dislodge shrapnel, Harken cut a small hole in the heart big enough to insert a finger to manually widen the valve.

Eleven years later, my father's heart was similarly accessed through an incision that began just above his belly button and ended just below his neck. I suspect ten-year-old Norman and eight-year-old Kathy were left with their grandparents. I imagine my grandmother Shirley sat in the waiting room rubbing her hands together the way she always did when she was nervous. My World War II veteran grandfather, Milton, although also nervous, remained stoic. My dad, twelve-year-old Billy, was tall for his age, but still looked like the child he was. Dressed in a hospital gown, he was wheeled on a gurney into surgery. Soon he became one of the hospital's very first successful open-heart-surgery patients.

The first procedure didn't work, although my father recovered well. So, in 1962, he had to undergo a second operation. This time, slight improvements were achieved, and the valve separating the right ventricle and the pulmonary artery could open a bit wider than it could before. All in all, the procedure was cautiously considered a success. Every ten years my father returned to Boston for stress tests and follow-ups. His health remained good. Meanwhile, he wore his scar, lifelong, with great pride.

———————

It was this one thing—this childhood surgery—that my father continued to come back to as he grew progressively sicker. For one thing, he had lived with that zipperlike red scar as a visual reminder of it for thirty-two years. Now the yellow fluid in his legs and lungs was increasing. The slight ankle swelling he'd shared with his mother and aunt rapidly surpassed theirs, spreading up his legs, filling his torso and, at times, his face and hands.

With each passing day, it seemed my dad was filling up faster and in more places. His decline was swift. Immediately following his first trip to the ER, he had canceled a family trip to Key West we were supposed to take during Hilary's Christmas vacation. My parents had gone to that Florida island several times, and my father had wanted to show it to us. We optimistically declared that the Key West trip was simply on hold. We would go when his health stabilized. Right then, we focused our energy on helping him to get better.

At forty-five years old, my dad stopped playing softball. He'd been a pitcher for years on his beloved amateur softball team, playing against other local businesses. He was still playing bass and singing with the band he'd put together with his friends, at practices and at the few small private shows they occasionally organized. But at their last show he'd looked so weak and gaunt behind his guitar, it no longer felt very rock and roll.

My father's primary-care physician was at a loss to explain what was plaguing my father. The wayward fluid indicated that something important was going wrong, but with each subsequent test, the answer continued to elude him.

My father was leaking lymphocytic exudate, chyle. The main problem is, if you are leaking protein and fat, you are not digesting them. Your body begins starving.

My father kept a medical journal. In it, he was careful and deliberate in trying to nail down the underlying cause for his ailments. In an entry from this time, he writes that he is "exhausted walking up a flight of stairs." He also notes that his worst-case scenario would be "an inability to gain weight."

Lymphedema was not, in and of itself, a diagnosis. Lymphedema, or the leakage of lymph from the lymphatics, is much more typically a symptom. Like a runny nose is a symptom of a cold, or like being out of breath is a symptom of running a mile as fast as you can. There are hardly any known conditions that involve the spontaneous leakage of chyle.

The lymphatic system is a part of our tissue, living outside of the vessels that make up our veins and arteries. The lymphatics collect or absorb whatever gets accidentally left behind as our blood courses through our blood vessels. Things like protein, random fluids, and even blood cells regularly slip out of our blood vessels as our blood flows. If the lymphatics don't collect them, we swell. If they can't collect them fast enough, we swell.

When people wake up in the morning with a full bladder, even if they haven't had a single thing to drink since they used the bathroom right before bed, it's because while they slept, the lymphatics had snatched up all that wayward fluid dropped by their blood vessels and directed it to their kidneys.

So my father's lymphedema (a term that means "lymph accumulation") should have been *caused* by *something*. But nothing seemed to be causing it. Even worse, when lymph *does* leak out of the lymphatic channels, for example in someone with heart or liver disease, there is no strong consensus in the medical community about *why*. And there aren't many ways to deal with it when it does leak.

Meanwhile, my father's heart was postsurgically near-perfect. The tissue was healthy. The beats were strong. There was no blockage in the thoracic duct, a main channel of the lymphatic system where digestion occurs between veins and lymphatic channels. And still, no one could find anything to explain any of his symptoms.

Another year passed and my father's swelling accelerated. The clinical term for his ailment went from "chronic lymphedema" to "massive ascites." Massive ascites is the result of chyle settling into the soft tissue of the abdominal cavity, surrounding the organs, and pushing out the gut to make room. My dad developed a hard and heavy bowling-ball-like gut that is common in people with extreme liver failure. He had breezed by a potential liver-failure diagnosis early on. A liver biopsy showed that his liver tissue was pristine for a man in his forties. As each possible cause of this extreme fluid leak was crossed off the list, my father was becoming more desperate.

My dad was the kind of compassionate doctor who always delivered scary news to his patients with a decidedly comforting addendum. "At least it's something we've seen before," he'd say. "The odds might not be great, but at least there *are* odds. Someone has survived this." His doctor could not offer our family the same soothing reassurance, and my father knew it. But somehow he remained steadfast, even at times upbeat. Having no diagnosis simply meant that we had to keep plugging away until we got one. If there was nothing new from where my dad stood in the midst of his current dire condition, maybe the answer was buried somewhere in his past. Maybe it was back with his twelve-year-old self, lying on a hospital gurney at Boston City Hospital with a finger opening up his heart valve.

My parents began taking trips to Boston, where I was newly enrolled as a freshman at nearby Tufts University. There my father began to see Dr. Michael Landzberg, a heart specialist who mostly works with children with congenital heart defects. Too old to be a patient at Children's Hospital, where Dr. Landzberg was on staff, my father sometimes checked into the nearby Brigham and Women's Hospital—affectionately called "the Brigham"—for testing. Other times he and my mother stayed at area hotels.

On the few trips my dad flew in alone for testing, Dr. Landzberg invited him to stay in his family's guest room. The relationship between my dad and his doctors often blurred the lines between friend and patient, in part because they shared a profession and a love of medicine.

Dr. Landzberg devoted himself to studying my father's case, realizing quickly that his illness didn't quite fit any known syndrome in the medical canon. He and my father agreed to aggressively seek answers. This meant asking my dad to undergo invasive procedures that as often as not weakened him but yielded few, if any, results.

With each passing visit, it became increasingly difficult for my dad to accept that maybe they would never find a diagnosis. His illness was getting worse and he was feeling worse. As it progressed, testing itself had become challenging.

One thing was growing increasingly clear, however. The problem wasn't his heart.

FOUR

Back in Boston for more tests, my dad was full of ports and plugs and holes. He and my mother had come to stay at the Children's Hospital Holiday Inn, which was the most depressing place in the world.

The thing about the Boston Children's Hospital Holiday Inn in 1994 was that it wasn't even trying to pretend it was *not* the most depressing place in the world. Nestled as it was between the high buildings of the Longwood Medical and Academic Area in Boston, not a speck of sunlight shone through its windows. The decor was brown and mauve with tan and gray-green overtones. The furniture and thick heavy curtains that covered views of brick walls bore the signature muddy tones of the 1970s as well as the flammable polyester of that era.

In one of these hotel rooms, a bath, of all things, almost killed my father. My mother filled up the tub for him and helped him lower his bruised and broken body into the warm soothing water. Methicillin-resistant Staphylococcus aureus poured into his open wounds and invaded his bloodstream. A staph infection. Those little party animals banded together and clotted the blood into all kinds of clever formations that even penicillin or its derivatives couldn't unravel. Overnight, he began running a fever as his body's vital signs steadily weakened. Dr. Landzberg had my father admitted to the Brigham, where he underwent surgery to try to contain the infection. But soon his fatigued body, swollen and raw, was failing in earnest. For the first time, my father was put into the ICU. His prognosis was dire.

My sister, uncle, aunt, grandmother, and grandfather all joined my mother and me in the hospital waiting room, where we began watching movies like *8 Seconds*—the Lane Frost story starring Luke Perry—and playing cards.

The room was perfect for little more than watching terrible movies. It wasn't the general-surgery waiting room—it was the ICU waiting room. It contained about fourteen chairs, three along each long wall, a love seat against the back wall, and two back-to-back rows of three chairs in the middle. The room was so small that only about three feet of aisle space wove a square around the middle chairs. The TV hung from the far corner ceiling, angled downward toward the desperate eyes of the room's pathetic and exhausted occupants. Sometimes, when all the chairs were full, I'd sit on the limited floor space. I'd lean back against Hilary's knees or rest my head on my mother's lap, hard-woven teal carpet making dents into my hands and, in warmer weather, the backs of my legs.

In my other life, in college, I had gotten my nose pierced and formed a band we called ethel g. because it was the nineties. We spelled our band's name in lower case because we read too much e. e. cummings. The guitar player once came with me to the hospital with his acoustic guitar so we could sing some of our songs for my dad in a private performance he tried to enjoy from his hospital bed.

But mostly I sat in that waiting room, where nurses sometimes delivered pudding or Jell-O and everyone held containers of coffee. I imagine some of the time we were waiting for my father to emerge from a specific procedure. Other times, we might have just been waiting for him to die. How strange to think if he had died during, say, the Lane Frost movie, we never would have known what was plaguing him. He would have simply passed away, a giant question mark looming over his head. Years would have gone by and we would have remembered our dad as a good dad who had probably picked up an ancient virus on a hike in the deserts of Israel. Or we might have wondered if he had broken something vital after his fall at the top of our steps.

My dad's unhappy forty-seventh birthday in February of 1994 reflected a life full of deferred answers. My parents spent the early part of the year visiting the Cleveland Clinic and the Mayo Clinic, as well as multiple hospitals in Boston and Columbus.

Then, one day an answer arrived. It wasn't neatly packaged like answers are on an episode of *House*. Its source had never gone to medical school, and usually left lipstick on your cheek after she kissed you hello. The source was my great-aunt Joanie.

Aunt Joanie was married to my great-uncle Nathan with whom she had two children. When Uncle Nathan died, Aunt Joanie remarried and had a third child. That marriage didn't last long. She took back her first husband's name, and also made it the last name of her youngest. Aunt Joanie remained firmly entrenched in our family. She is known for always being one of the first to RSVP yes to family bar and bat mitzvahs and weddings, so it wasn't surprising that she drove from Yonkers to Boston to visit my father in the hospital. The moment she saw him, she recognized his symptoms. In 1961, Aunt Joanie's thirty-four-year-old husband, Nathan, died of a three-year illness. She immediately told my father that he had the disease that killed his uncle Nathan. Whether it was because he didn't believe her or because deep down he already knew, he seemed to ignore the information at first.

But Aunt Joanie didn't give up. Although the early winter of 1994 was cold and gray, my aunt got into her little brown Honda and drove herself straight to the National Institutes of Health in Bethesda to collect Uncle Nathan's now-dusty chart. Her husband had spent close to a year in a hospital bed in the NIH. She knew it well. She had made the journey dozens of times, both alone and sometimes with her small children. The chart wasn't easy to find. Aunt Joanie had to go back and forth two more times before they handed it over. She drove it to Boston.

My dad was fighting through another near-fatal infection. Lymphedema makes skin exceedingly prone to infection, since it reduces blood flow to the affected areas; in extreme cases, when the patients are undernourished, the situation is even worse. My dad did not feel well the day his aunt arrived clutching a three-ring binder that contained all the unpleasant details of her young husband's death.

He was fourteen when his uncle Nathan died. According to some of my father's cousins, Nathan had looked terrifying in his dying state. One of his irises had acutely dilated, which caused the lid to droop and finally seal shut. His body swelled with fluid. Late in his illness, Nathan's small head sat atop a giant, fluid-filled body. He was exhausted and in excruciating pain. All who saw him remembered this time as a nightmare.

So I could wager a guess that looking at Uncle Nathan's chart was about the last thing my father wanted to do—especially as his concerned aunt pointed to it, saying, "See? Just like Nathan . . ."

While she was presenting the chart, someone said, "Tell Joanie to wait in the hall." So I sat with Aunt Joanie in the hall that day, and she handed me the binder. She was wringing her hands. She felt terrible. I opened the chart to the first page and began to read.

The parallels were impossible to miss. Chylous ascites—another name for the lymph settled in the gut—and lymphedema, the swelling in the appendages, all "of unknown origin."

Unlike the rest of my family, I felt excited reading it. The words described my father's condition almost to a T. That someone else had had it seemed like a good thing to me. Here at least was *some* information rather than more of the *non*information that doctors had been giving us. It had never occurred to me how lonely my father's illness was until that very moment when we were no longer alone.

My dad and my grandparents received the news coolly. I imagine that their memories of Nathan's struggles—the isolation and the endless misdiagnoses, the confusion at the end of his short life—overwhelmed them. I imagine that the idea that this condition was hereditary only exacerbated their fears.

Several weeks later, a second medical chart found its way into the thick of my father's medical conundrum: his grandmother's. Uncle Nathan's mother, Mae.

Although Mae's 1955 medical chart was something of a relic when viewed side by side with her grandson's from 1994, it made one thing certain: the condition my father was fighting to survive had already been seen at least twice before.

Upon realizing that you are going to die—truly internalizing that your physician has just said, "There's nothing more we can do"—you will likely go through five or so stages of grief until at last you settle into a profound acceptance: now you know how death will come. At least you know the outline of it, the shape it's going to take. It will look like this cancer or that failing heart or this gene.

You might handle this news in one of three basic ways. You might ignore it and embrace your denial. In fact, this is the most natural response to death because this is how most of us live our lives. In our day-to-day existence we might occasionally nod and say, "Sure, sure, I know I'm going to die," but we don't regard it as fact. It's less specific than that. Maybe we think beyond it—*I will go to heaven*. Or

we think around it—*What will the world be like in the year 3000?* But
we don't truly sit with the reality that the death of every single one
of us is imminent. When you receive the information revealing *how
you will die,* there is no reason anything should become any different than you've always known it to be. You will go to sleep, you will
wake up, you will go to work, and you will wonder if maybe you are
supposed to be doing something differently.

But what? What is different about today? Especially if you still
feel relatively healthy, why would knowing *how you will die* suddenly
prompt you to take a trip to China or start auditioning for parts in
movies? Even with this new knowledge, your life continues to just be
your life. At worst, you might take up heavy drinking, or acquiesce
to taking those opioid painkillers your doctor is offering; you might
sleep more or watch more TV. But all of these accomplish the same
goal—ignoring your impending mortality.

The second way you might handle this news is that you might
begin frantically assessing where you stand in your search for meaning. You might begin by making some "I love you" or "I'm sorry"
phone calls. You might try to be more present as you sit beside your
children watching TV, or hold your husband in bed at night. You
might read a few books about God, learn to salsa dance, or organize
your files.

Or, a very human reaction to knowing you are going to die, when
it suddenly becomes for realsies, is that you might stand up and go
to fucking war. This will be a new kind of war. It's actually 98 percent
bullshit and 2 percent fantastical. That's okay, because you don't need
fact, reality, or the Second Amendment. You need miracles and Pa-
leo diets, crystals and healing hands. You watch movies where a tu-
mor magically disappears. You follow stories of heretofore unknown
treatments. You call universities. You seek out Katie Couric or Ellen DeGeneres to tell the world about it. You gather up every tool,
magic wand, and superhero cape you can find and you stand up and
fight. It is at once the most rational and the most insane you will ever

be. It will bring the most satisfaction and the most heartbreak you will ever endure. You will get the most done in the shortest amount of time, and you will almost certainly die anyway. But you will go down fighting, and everyone around you will call you valiant at your funeral. (If you are watching that funeral from Beyond, you will be flipping them off for saying it.)

Ideally, whether you have that death diagnosis yet or if you are still blissfully disregarding your certain demise, you should diligently practice all three of the above. You should fight to live long and well, you should experience meaningfulness as often as you can, and the rest of the time you should just ignore death completely.

When word was given to my father that his condition was so rare he was possibly only its second or third documented case, he plunged himself more deeply than ever into war mode, but medical genetics wasn't ready for his case yet. In 1994, very few people were capable of fighting *genes*. There were too few tools. We knew so little. I remember feeling suddenly invigorated. I knew next to nothing about genetics beyond what I had learned in a tenth-grade science class. In the early 1990s, though, it seemed like a field that was developing at breathtaking speed. I wondered if maybe my dad's case would suddenly blow the field of genetics wide open, now that we knew what we knew. It was clear my great-uncle had sought answers defiantly until his dying day. But that was the 1960s. This was a million years later. We had CDs. We had over fifty television channels. This was the future.

I felt certain we could use the little that Nathan's doctors—and even his mother Mae's—had learned and analyze it in the light of today's medical innovations. Three cases could certainly tell us all we needed to know in order to save my father.

My dad wanted to agree with me, but looking at those charts, digging through his family's horrific past, meant facing fears he hadn't yet faced. As a doctor, he had a better chance of uncovering the latest available medical practices, but genetic treatments were hardly mainstream in the mid-1990s. His cynicism was earned. It's possible that on some level, he believed that the jig was up—that maybe it was time to send out baked goods to people he'd pissed off, or needlepoint a meaningful aphorism on a pillowcase. Even if my father wasn't ready to face that kind of inevitable, he understood something crucial that a lot of people in the mainstream still didn't completely comprehend, and that frankly I hadn't really even considered: if this was genetic, then he might not only be fighting for his own life.

He might be fighting for my life too.

FIVE

In the mid-nineteenth century, Gregor Mendel, an Austrian monk living in the Czech city of Brno, moonlighted as a botanist. He was a round, balding, wire-rimmed-glasses-wearing friar who plodded around a plot of land behind a small parish in the Moravian countryside. He'd grown up on a farm with two sisters and was so desperate to study that, despite a possible lack of interest in divine providence, he joined the Church because it offered a free education. As a monk, Mendel began to tend a garden where he grew pea shoots, among other things. He became curious about the varied colors of pea-shoot flowers. What made one plant's purple and another's white? Mendel began to obsessively watch and record how this trait in particular transferred from one generation to the next. Thirty years later, his research posthumously made him the father of genetics.

There aren't many good stories about Mendel. He was a man of the cloth who watched and recorded the growth of pea shoots. In addition, he wasn't too great of a test taker, managing to fail his teaching exams twice, which limited his opportunities to share his wealth of knowledge in his lifetime. In fact, Charles Darwin, a contemporary who allegedly possessed a copy of Mendel's breakthrough study, *Experiments in Plant Hybridization,* allowed it to languish unread.

Mendel's laws of inheritance or Mendelian inheritance are composed of two theories that became the basic rules of genetics. This boring, exam-failing friar first took two pea-shoot plants he'd bred and grown himself, one purebred with white flowers and the other purebred with purple flowers, and mated them. Would the resulting plant have petals swirled with purple and white? Would they be a light purply mixture of both?

No. When Mendel bred the purple-flowered plant with the white-flowered plant, the flowers of the second-generation plant were all the very same rich purple as *one* of its two parent plants.

When Mendel took two of these second-generation purple-flowered pea shoots—born of one white-flowered and one purple-flowered parent—and bred *them* together, he made a powerful discovery: one out of four of their offspring yielded a white-flowered plant. The other three had purple flowers.

Mendel's theory was simple. Purple flowers were a "dominant" trait because only one parent had to share the information for purple flowers with its offspring for the trait of purple flowers to manifest. Meanwhile, white flowers were "recessive" because both parents had to pass the information for white flowers in order for their offspring to yield white flowers.

The second theory suggested that this white/purple information was stored in heredity units later called "genes," and that each of us had two of them for each trait, one from each parent, which later became known as "alleles."

Mendel's ideas led to a scientific breakthrough. By all accounts, his success was wholly deserved. The young scientist had been so desperate to study that even at his most impoverished he refused to give up his research. When things were at their worst, one of his sisters gave him her dowry money in order to keep him fed. Never forgetting his debt, Mendel later put all three of her sons through college and guided them toward promising careers in the sciences.

As this new area of study expanded, Mendel became one of its earliest luminaries. Those who shared his work with the world lauded it for enriching their own studies. The great botanist, Hugo de Vries, who was the first to suggest the concept of "genes" and "mutations," conceded that it was only after studying Mendel's work that he truly understood his own.

By the 1950s, the American James Watson and the Brit Francis Crick had joined forces at Cambridge University and were awarded the Nobel Prize in Medicine in 1962 after becoming the first to unlock the mysteries of the structure of DNA.

They used the work of a brilliant young X-ray crystallographer and chemist named Rosalind Franklin. Franklin had produced one of the clearest images of DNA ever developed by shining a light through a single crystal of DNA and then photographing the re-fracted light. The DNA crystal itself, however, was much too small to see using the available equipment of the 1950s; she could only discern the light refracted onto the wall. Franklin was working with the shadow lights of DNA. She had to obtain the clearest picture of those dancing lights so that molecular biologists like Watson and Crick could work backward from that image and figure out the shape of the crystal that had caused them. Franklin managed to produce the clearest, cleanest picture of one of those shadows ever taken.

With Franklin's images, Watson and Crick were able to prove that DNA looks like a twisting ladder called a "double helix." The double helix of DNA is composed of four different bases, or nucleo-tides. They line up, one after the other, over and over, all the while pairing off with the same buddy, over and over.

These nucleotides are not rungs on the ladder of the double helix, as you might imagine. They are in fact spots running up each of the two side-by-side rails of the ladder—with one nucleotide sitting on, say, the left rail of the ladder with its partner on the opposite rail. The "rung" is made out of those two buddy nucleotides "holding hands." A series of these pairs (called "base pairs") makes up a gene. The DNA coils, uncoils, and with the help of RNA becomes a thick roll of DNA—the blueprints of life. Each thick roll of DNA is a chromosome. Another way you might imagine a chromosome is by picturing a Twizzler candy rope (or a Red Vine, depending on your coast) as a bunch of DNA. A single section of that red rope is a gene. The entire rope coiled together is a chromosome. So, a "chromosome" is a cluster of genes that make up the stick of licorice. Humans generally have twenty-three pairs of chromosomes, inheriting one of each pair from each parent.

DNA—with all of its base pairs that make genes, and genes that make chromosomes—lives in essentially every single cell in a person, a fish, a tulip, a bacterium, and all other living things. The DNA in each cell is always identical to the DNA in every other cell of that thing. The job of DNA is to be informative. It doesn't have to be a snappy dresser or dance well. It's the stodgy egghead of the cell, simply being the rules and regulations of the "you" the cell is making. It doesn't actually *do* anything. It isn't creative or wily. It just makes copies of itself and makes something called RNA—which differs from DNA only slightly in that RNA copies *itself* and proteins. Something called a ribosome reads the RNA made by DNA and then does all the interesting stuff—like making the human. DNA is essentially an instruction manual that the RNA reads out loud to a ribosome that does the stringing together of amino acids to build the thing the instruction manual DNA relays. When you think about it, DNA is so fundamental that the next logical questions are: Who told the DNA the rules? Who wrote the manual?

When a *genome*, or every letter of every line of DNA, is written just so, science has learned that people can live long, strong, and healthy lives. If a letter or two or a thousand are written incorrectly, however, it is considered a "mutation," or as today's scientists typically refer to it, a "variant." Sometimes absolutely nothing ever comes of these misprints. Other times, even one single miscopied, changed, or eliminated letter can mean the difference between health and sickness, life and death.

The double-helix revelation was astonishing, a revolutionary breakthrough in the scientific community. Suddenly we could "see" DNA. We could actually "see" our genes! Now that we could see them, like a ribosome, we could begin to read them. And read them we did. One hundred years after Mendel's purple and white pea shoots, hundreds of scientists of all different races, genders, and nationalities worked together to unmask the human genome. In a worldwide project, they laid out in exhaustive detail all the roughly 20,500 genes that make up a human being. Some say it is the greatest scientific accomplishment of all time.

In 1994, while my father was spending time battling infection, ignoring Aunt Joanie, and looking for answers at Brigham and Women's Hospital in Boston, the Human Genome Project was still seven long years away from publication. Genetics has always been a science of patient observation over time. Studying generations across plant lifetimes can sometimes be accomplished in a matter of months. In humans, the same studies can take years—lifetimes. Unfortunately, as anyone with a genetic illness will tell you, and what my father was trying to forget, was that "lifetimes" were exactly what he didn't have.

SIX

Soon after Aunt Joanie brought Uncle Nathan's chart to the hospital, Dr. Christine "Kricket" Seidman, one of the world's leading genetic researchers, came onto my father's case. Dr. Landzberg, who had taken my father on as a houseguest during his early Boston visits, might have alerted her to my dad's case, or a cousin who says her neighbor knew Dr. Seidman from medical school might have told her about him. Regardless, when I met her, I remember thinking that she didn't seem like a doctor. She was far too easy to talk to, and much too pretty. She had sparkling blue eyes framed by short black hair, and she told us—no, *insisted*—that all of us call her "Kricket." I've come around to calling her "Dr. Kricket."

Dr. Kricket, a cardiologist by training, has spent most of her career looking at genetic conditions of the heart. Initially, she found my father's case interesting because of the cardiac component—his sticky valve from childhood—but she was also drawn to my parents. She found their situation heartbreaking and admired their steadfastness in trying to figure things out.

Lucky for us, Dr. Kricket already had her own thriving genetics lab. There, she and her husband, microbiologist Jon Seidman, had spent impressive careers uncovering genetic links to a multitude of congenital heart disorders. They received wide acclaim for their breakthrough studies of familial hypertrophic cardiomyopathy (HCM), a condition best known for taking the lives of young athletes on sports fields. It predisposes sufferers to a thickening of the heart tissue, which can lead to sudden death. The Seidmans mapped that variant and then spent years studying it.

Very early on, it was clear that my father's murmur, while somehow connected to some genetic piece of pie, was probably not central to his strange and harrowing illness. Because his illness did not seem to involve a cardiac component, he was not going to be a typical case for the Seidmans.

Nonetheless, Dr. Kricket made the decision to stay with our family for the duration. Her and her husband's research lab specialized in figuring out the genetic reasons behind certain conditions. Once she had begun working on the case, she wouldn't leave it, regardless of any odd, remote, or even erroneous paths it took her down.

Dr. Kricket and her team began their investigations of rare diseases by constructing a family narrative. She would compile stories and pictures as well as medical data to pull together a cohesive history to study. In that spirit, I decided to find out what I could about Dr. Kricket's own family narrative.

She grew up the middle child in a family who didn't produce doctors, but did produce a lot of illness. She noticed that the one person around her who wasn't sick was always the doctor, so she decided she wanted to become one when she grew up.

She got into Harvard for undergrad and followed a traditional premedical course load, which meant starting with an introduction-to-biology class her freshman year. Jon Seidman, also premed, took the same class. Soon the two were in love. Kricket came from a traditional family and understood that if she was going to move somewhere with a man, she was going to do it married. When Jon had an opportunity to perform doctoral work at the University of Wisconsin, they decided to tie the knot.

By 1984, they were working together and became two of the first scientists to successfully clone a piece of heart tissue. Dr. Kricket completed her residency at Johns Hopkins, where she turned her focus to research. Mentored by Victor McKusick, known as the father of medical genetics, she lent her drive and genius to the Human Genome Project. Twenty-five years later, she called on her mentor one more time. He was now in his eighties but still practicing on a limited basis. At that point, Dr. Kricket had met a patient at Brigham and Women's Hospital. He was a forty-seven-year-old man who seemed to have been born with a congenital heart defect and was now riddled with wayward lymphatic fluid throughout his body.

After she took on my father's case in the mid-1990s, Dr. Kricket's primary job was to figure out what other genetic illnesses showed similar symptoms. Her goal was to find a family or group of families with physical presentations that mirrored my father's. This was the part of the plan I liked the best. Once we found that group, we would pool our collective knowledge, build a community, maybe start a support group, and have fund-raising bake sales, dance-a-thons, and ice-bucket challenges.

———————

I was thrilled when I heard that Dr. McKusick was going to work on our case. In the world of genetics, he was something of a celebrity. A cardiologist by training, he'd chosen to study the heart because he loved the musicality of heartbeats. A hereditary heart condition called Marfan's syndrome led him toward a developing field that later became known as medical genetics, as the head of

Johns Hopkins brand-new Division of Medical Genetics in 1957. In 1969, he was one of the first to propose a human genome map, which many believe led directly to the Human Genome Project— first published in 2001. It wasn't until 1983 that the first genetic illness was "mapped," pinpointing an exact genetic variant inside the complex proteins of DNA coding.

I recently read a popular book by a journalist named Rebecca Skloot called *The Immortal Life of Henrietta Lacks* that principally focused on the history of immortalized cells. Immortalizing cells has been an enormous boon to genetic medicine and research. It protects, purifies, and safeguards specific cells for research, sometimes long after the host has died. More specifically, immortalizing cells allows researchers to keep a single cell alive in a frozen environment for as long as it is required for study. Doctors were soon capable of cloning cells an infinite number of times.

Henrietta Lacks was diagnosed with cervical cancer in 1951. Unbeknownst to her or her family, doctors harvested some of her cancer cells and preserved them. Because cervical cancer cells are already structurally heartier than other cancer cells, Lacks was a good candidate for an experiment to see whether or not her cells could withstand the then-crude conditions to become immortal, a complex process of heating, cloning, and freezing them. It worked. For the first time, cells replicated through cell division outside of Lacks's own body. Eventually, Lacks became the first person to have her very own living cells outlive her. If you consider that Henrietta Lacks died in 1951, her still-living cells have outlived her by more than sixty-five years.

Called HeLa cells (the first two letters of her first and last name), Lacks's cells have helped the scientific community to accomplish wonders, like discovering successful cancer treatments and studying how a body might survive space travel. Lacks's contribution to science, however, remains a controversial one. First, neither she nor her family gave consent before her cells were harvested. Largely, this is a reality for all of us when we enter a hospital. Our tissues can be discarded or utilized at the hospital's discretion. That's why you rarely hear about people taking home a souvenir appendix. Second, many people ultimately made a lot of money on the results of research conducted using HeLa cells, while the Lacks family languished in relative poverty in urban Baltimore.

For better or worse, HeLa cells served as an early and vital tool for Dr. McKusick in his ongoing development of medical genetics. Skloot writes about Dr. McKusick in her book. Today, medical genetics, by way of immortalized cells, is directly responsible for many of the innovative medical weapons that will one day cure cancer and fix many of our worst genetic anomalies. DNA lives in cells, so the same gene in the same cell can keep on living, even after the host of that cell has died.

Today, scientists immortalize all kinds of human tissues, which is important for people like Dr. Kricket who are studying families with genetic conditions. Specimens for her tiny study of my father slowly began to fill up a freezer in a single corner of her sprawling lab. After all, when she figured out what my father was suffering from, she'd want to be able to share those specimens, whether or not she was in time to save his life.

———————

Only so many illnesses exist in the world. In 2007, the World Health Organization (WHO) identified 12,420 human diseases. However, in 2004, GenBank—a DNA-sequencing bank and online forum—had already categorized 22,000 disease categories based on DNA makeup alone. Today, the WHO reports around 10,000 named monogenetic (single gene) disorders. Given the rapid growth of genomic medicine, and the formidable ease and low cost of genome mapping, that number is certain to grow. Scientists will continue to find genetic nuances among illnesses once forced into a single box.

For example, "cancers" are now being broken down by genetic makeup so that they may be treated uniquely. Unlike my family's gene variant, passed from parent to child via egg or sperm, cancer mutates one cell at a time until the number of cancer cells begins overwhelming the body's systems. A growing field of medicine called "genomics" studies the genome of both the cancer *and the patient.* This vital step in cancer treatment is changing the way medicine is practiced. Instead of looking to statistics culled from a pool of out-dated data, doctors can narrow their scope down to a specific type of cancer and look to those treatments that have the best possible outcomes. Who knows? One day the blanket term "cancer" might even become obsolete as genetic distinctions highlight various cancers' profound differences.

But in the mid-1990s, when my dad was sick, the medical community only knew about roughly 741 genes linked to genetic disease. Other diseases that were believed to be genetic had not yet been linked to specific genetic anomalies, and those that had been linked numbered only in the dozens. The possibility that Dr. Kricket was looking at a whole new disease bordered on the nearly impossible. It just didn't happen, at least not in plain sight, not the way this illness was behaving and playing out among three generations. However, no other populations seemed to share, or ever have shared, my father's symptoms.

Discovering a new disease is always unusual and surprising. Genetic diseases do not pass from host to host by way of germs like the measles or the flu. A genetic condition must either occur in a person who can live with it until he or she reaches an age when they can pass it to offspring, like Huntington's disease, or it has to lie dormant, recessive in healthy parents, until it finds a match that will produce a child with the condition. Both conditions illustrate how genetic diseases that impact children continue to pass among generations by "carriers." Carriers are people who do not experience the detrimental symptoms of a genetic condition, but still carry the gene in half of the chromosomes in their cells. They have a 50 percent chance of passing that particular gene to their offspring. Like Mendel's second-generation purple-pea-shoot plants, that gene for white flowers was there; it was simply dormant and not being expressed.

Genetic diseases with an adult onset like my family's are very rare, but they are the ones that predispose—usually guarantee—that people will develop grave symptoms later in life, as my father did. In the case of my family gene, based on the three cases we knew of, it seemed that the disease affected its victims only during or after their fertile years.

Humanity probably never knew about some genetic diseases. Perhaps the detrimental genetic anomaly simply naturally selected out of existence, or died out because it affected a single life, a baby or a child, who didn't survive long enough to reproduce. But if Dr. Kricket figured out that my father was truly alone in this, if my father was only the third person in three generations to have a unique genetic condition, then what she was looking at was a private mutation—a mutation impacting a group of related individuals that, over time, could expand into something called a *"founder population."* This is a group of distantly related people exhibiting the symptoms of a heretofore unknown mutation—a *"founder mutation."*

The founder mutation of Tay-Sachs disease, a rare, brutal child-killing illness most common among Ashkenazi Jewish populations, originated thirty to fifty generations ago, dating all the way back to the eighth or ninth century. Cystic fibrosis, a respiratory condition that affects hundreds of thousands of people of mostly European descent, is even older, originating in Europe fifty-two thousand years ago, or 2,625 generations. Sickle cell anemia, a debilitating blood disorder of principal significance in sub-Saharan Africa and among African Americans, is comparatively new, but still affects more than ninety thousand children per year. The disease is believed to go back to a single common ancestor who lived a mere five thousand years ago.

My father's disease remained uncategorized, but as Dr. Kricket was learning, it bore striking similarities to stories she was told about his grandmother and uncle's diseases. She couldn't make a definitive claim about whether or not my father's illness was genetic, and what it might mean if it was, but she was prepared to find out. First, she needed tests, she needed blood, she needed to preserve some of his cells, and above all else, she needed stories, and lab work from all the members of our family.

―――――――――

Dr. Kricket contacted everyone she could find who was related to my great-grandmother Mae. She recommended testing or offered to fly them in to Boston, one by one. She put family members up in hotels and gave us spending money for food. One weekend in 1995, I drove from Tufts, where I was a sophomore, to Logan Airport to pick up my sister and our cousin Danny.

On that trip, Dr. Kricket took our vitals and focused on our hearts. She was going on a hunch based on my father's childhood heart surgery, lifelong heart murmur, and little else. The tech listened to our heartbeats closely. The gel on my chest was cold and sometimes the angle of the ultrasound wand dug into the soft spots between my ribs. After a few minutes of carefully tweaking the echocardio-gram, which provides images of the heart, she turned up the sound. It wasn't immediately obvious. I mostly heard whirring noises, but Dr. Kricket was nodding. She pointed to a cluster of colors beside me on the monitor. Then I could hear it and even see it: a tiny but distinct little whoosh of air in the middle of my heart thump, which according to Dr. Kricket was caused by turbulence as blood exited my heart to the vessels. *Buh-sh-bum. Buh-sh-bum.* A murmur. Hilary had it too.

Danny, my aunt Kathy's son, didn't have it. We high-fived him. The two lists divided, as cousins, aunts, uncles, great-aunts, and uncles all flew in or went to their own family doctors for a checkup. Some, like Danny and his mom—my father's little sister—didn't have the murmur. My dad's younger brother, Norman, who lived in California, had visited a lab at the University of Southern California with explicit instructions from Dr. Kricket. His doctors didn't detect a murmur either, nor was there a murmur in the hearts of his children.

Hilary and I were placed on the other list, the one composed of those in whom Dr. Kricket *had* found murmurs. In all, out of forty-one of Mae's direct descendants—five children, eleven grandchildren, and twenty-five great-grandchildren, including Mae and her son Nathan—thirteen of us had the heart murmur that provided Dr. Kricket her very first clue.

She put out a call looking for similar cases in the medical community. She contacted descendants of Mae's two brothers, and any other distant relatives she could find. She pored over literature. She consulted with Victor McKusick. But she found nothing.

SEVEN

As my father lay swollen and dying, and as Aunt Joanie stood clutching the chart of her dead husband, Nathan, I remember asking my grandmother Shirley again what she remembered about the death of her mother. Great-Grandma Mae's death was shrouded in mystery.

In the past, my grandmother had shrugged her shoulders and answered, "I don't know." This time, as she sat near the hospital bed of her very ill son, she replied bleakly, "What do you think? She filled with water."

But the story Dr. Kricket collected from my grandmother and her siblings was decidedly better drawn.

Mamie Bloom stood an inch over five feet tall. The oldest child in her family of three children and the only girl, she met a young rabbi through a friend. Rabbi Morris was Pittsfield's first Orthodox rabbi. He had fought to make ends meet as a seventeen-year-old Russian immigrant. While he taught himself English and put himself through Yeshiva University in New York City, Rabbi Morris was the kind of poor that meant sometimes you just didn't eat. His mother and siblings helped him when they could, but eventually his strong faith and an even stronger work ethic landed him his own congregation in the foothills of the Berkshire Mountains. He was young, tireless, and deeply religious. The second youngest of eight children, Rabbi Morris stood five feet four inches, with a handsome face and a stylish mustache. A fated fix-up by his older sister with an American-born girl with a large and ready smile who lived four hours away in Brooklyn changed his life.

Mamie, he decided, would make a good wife to bring home to Pittsfield. There was little in the way of courtship—a few letters, a few visits—but soon the two became engaged.

Mamie and Rabbi Morris were a part of a new generation in a new world, where people were no longer forced into arranged marriages like their parents. With her fiancé's blessing, Mamie changed her name to Mae, which she felt sounded more dignified and better suited to the fiancée of a rabbi. She and Rabbi Morris were married in her hometown of Brooklyn. She followed her husband to Pittsfield, where she took her place in the congregation as the rabbi's wife—a position of honor.

Their early years were happy ones. Mae's first child was a girl. The birth certificate read Sarah Beila, but they called her Shirley. A year later, Norma was born, followed by Nathan and Harry. During one of those pregnancies, Mae's ankles began to swell, although for her, this was little cause for concern. Her mother's ankles had also become swollen during a pregnancy, and had remained swollen ever after.

Mae gave birth to her last child, Joseph, whom everyone called Yussy, ten years after Shirley. Mae doted on her children. In her household, she followed her husband's many rules and customs. She kept kosher, she dressed in compliance with the customary modest code, she disciplined her children, and she made their clothing so that they were admired, sitting well behaved in a row every Friday night and again Saturday morning at the well-attended services led by Rabbi Morris. No one at those services could look on that family without agreeing that they were blessed.

I like to picture Mae holding baby Yussy. A ten-year-old Shirley with a bow in her hair, glasses on her pretty face just above the two pronounced dimples in her cheeks. She's kicking her nine-year-old sister, Norma, for pinching her. Norma, Shirley thinks, is the pretty one. She has big bright blue eyes and dark brown curls. They are sitting with the women in the noisy balcony seats. Chatter here is normal, above the humming and monotonal mumbling of the prayers being recited below. Nathan and Harry are down with the men because their father wishes them to be there, although they are only eight and six years old, respectively. Harry is bored but Nathan knows to shush him or take him back to their mother if he becomes too restless. They are quiet boys most of the time, and obedient, singing along with the prayers that, despite their age, are familiar to them from years of study and repetition. The family walks home together, or perhaps Rabbi Morris comes along later. They are laughing, skipping, and playing, spending that Saturday, like every Saturday, together.

———————

I didn't know Mae. Her eldest child was my grandmother Shirley, whom I knew very well. I called her every day and went to visit her twice a year in Phoenix. When she could, she came to visit me. She told me once that Mae was already the mother of five children the morning her right arm doubled in girth from hand to shoulder.

Shirley was only twelve the day her mother's arm swelled. When I pressed her about her mother's illness, she could easily recall the swollen arm, but not the minutiae of what came after. Because Mae probably just wrapped her arm in a cold towel without much fuss, there may not have been much to remember.

Mae's swollen arm, unbeknownst to anyone, signaled the beginning of a harrowing, if slow-moving, fight for her life as her health mysteriously began to deteriorate. Her doctors did what little they could to ease her earliest symptoms. Her swollen arm was a nuisance and a physical disfigurement, but it wasn't life-threatening. One doctor sent her home with a manual pump she'd wrap around her arm, utilizing air pressure from the pump like a massage to keep the fluid loose. Mae used it for many years before she went to bed at night.

Meanwhile, Mae's other tests continued to come back normal. But when, three years later, her breathing became labored due to pleural effusions—leakage of fluid, much like in my father's case, into her lungs—all bets were off. Mae was thirty-eight years old. Her children ranged in age from four to fourteen. Not being able to breathe wasn't an option. At the hospital, doctors were able to successfully drain off some of the fluid that had settled into her chest, taking some of the pressure off her lungs. This manual procedure, which predated the mechanized tapping that was used on my father, was also called a "thoracentesis" or "pleural tap." In the 1930s, when Mae first underwent it, the procedure required a small incision between the ribs through which a rubber hose was inserted into the chest wall, where it channeled out the fluid under the force of its own pressure. By also employing manual pumps, the doctors slowly drained out the fluid.

The first time Mae was tapped, she watched as an alarmed and confused medical team extracted about fifty milliliters of lemon-meringue-yellow lymphatic exudate—or chyle—from her chest. No one knew where the fluid had come from, but Mae's doctors did charge forward with a few ideas about what might be causing her labored breathing. Rudimentary treatments for breathing conditions pioneered during the Great Depression typically had only limited success in helping to ease discomfort, but protected the general population from catching some of the viral and germ-based respiratory illnesses of the day, most typically tuberculosis, an infectious bacterial disease usually occurring in the lungs. Small quarantined colonies for the infirm were set up on hospital grounds. "Fresh air" was the unofficial prescription. Hordes of sick people were often herded into large tents and ordered to simply inhale and exhale.

At this point, Mae had been officially diagnosed with pleurisy, a condition in which a series of membranes covering the lungs, called "pleura," become inflamed. It is a symptom of tuberculosis (TB), which in the 1930s often proved fatal. By 1938, doctors at Mount Sinai Hospital in New York had tested Mae's fluid for bacteria and ruled out TB. What they knew was that her lungs were filling with chyle, and they had to do something to drain it off.

Because Mae's symptoms did not point to TB—the swelling in her limbs, the swollen belly, and the breathing complications that were not in the least bit contagious—her condition did not fit any known malady. To treat her, the doctors decided to follow the same protocol they used for other breathing disorders of the day: Mae would sleep outside and live in relative seclusion.

One day two years after she woke up with a swollen arm, Mae and her husband took their oldest child, fourteen-year-old Shirley, for a walk, leaving the other kids at home. They told Shirley that her mother was going to go away for a while to a hospital to help her breathe.

Mae was sent to Coolidge Memorial Hospital on the southwest side of Pittsfield, twenty miles from the family's home. Patients with tuberculosis, contagious and otherwise, lived in tents about half a mile downhill from the main building, where Mae took up sleeping quarters on a beautiful wraparound screened-in porch with three to

four other patients, all TB patients undergoing treatment.* There, she
was cared for by a kind staff.

———————

Mae spent six months convalescing on that porch, followed by years
in and out of Mount Sinai Hospital in New York, a hundred and
fifty miles away from her home and family. Meanwhile, Shirley
grew into the caretaker of her household. She cooked, which she
says came naturally—"I was born just knowing"—helped to raise
her three younger brothers, and fought for power with her one-year-
younger sister.

At the time of Mae's stays at the hospital, her family did not own
a car and relied on friends and neighbors to take them to see her.
Her children were young when her infirmary stays began—Shirley,
fourteen; Norma, thirteen; Nathan, eleven; Harry, nine; and Yussy,
four. Their visits were infrequent and rarely long enough.

When Mae was released from Coolidge Memorial, her condition
had stabilized. She began visiting Mount Sinai Hospital in New
York City, undergoing exploratory procedures to seek out the cause
of the fluid that was steadily, if slowly, increasing. She stayed for
weeks at a time with her parents in Brooklyn, a four-hour journey by
train from the Berkshires. On several of these occasions she brought
Yussy, her youngest. Mostly Yussy was left in the care of his grandmother or one of his sisters.

Ten years after the appearance of her first symptoms, when her
oldest children were married with children of their own, Mae began
undergoing radiation treatments in a "we'll try everything even if we
don't know what we're treating" sort of spirit. Radiation treatments

———————

* TB patients were believed to no longer be infectious after about two weeks of steady
treatment and were usually released, although we now know this information was
erroneous.

were the conventional recommendation for many life-threatening illnesses. In order to receive the treatment, Mae walked to Pittsfield General Hospital, often with Rabbi Morris, Shirley, or Norma, who helped out on her visits home from New Jersey. After the treatments, Mae usually threw up in the street—a side effect of the radiation.

By 1955, nearly twenty years after Mae woke up to that swollen right arm, the accumulation of fluid in her body had become unbearable. She underwent repeated tapping procedures to extract as much of it as possible, but by midsummer, the tissue around her lungs had become so scarred that the fluid was unreachable. Shirley, by now a mother of three and living with her in-laws, helped out when she could. When her father instructed her to accompany her mother to see yet another specialist, this time across the state in Boston, she reluctantly agreed, feeling as she did that it was her father's responsibility to accompany his wife to the appointment. But Shirley was not the only family member whom Mae's protracted illness was making into a martyr.

Decisions about Mae's treatment were left to a reluctant Shirley. A doctor at the Pratt Diagnostic Center wanted to put her broken body through a deeply invasive procedure called a "thoracotomy." He told Mae and Shirley that her chance for survival was fifty-fifty. A thoracotomy involves a chest and abdominal incision that opens up the skin, muscle, and rib cage in order to gain access to the patient's heart and lungs. Cutting open skin that is riddled with lymph is always dangerous. Even a small, everyday cut can easily become infected, so larger, more prominent incisions risk deadly infections. In a woman whose body was already in a weakened state, it's likely Mae's chances were far worse than fifty-fifty. But the Pratt doctor was the only one with a plan. And Mae couldn't breathe, so she couldn't afford to be choosy.

Shirley gave the okay for the operation, which was performed at Albany Hospital on August 22, 1955. Her father joined her in the hospital waiting room during the five-hour procedure. Immediately following the surgery, the incision the doctor had made began to fester. Over the next two weeks, Mae's condition remained touch and go. Finally, on September 9, 1955, she died. The infected thoracotomy scar on the left side of her chest noted in the report was listed as the primary cause of death—although the gross quantities of lymphedema in every region of her body were cited as a contributing factor. She was fifty-five years old.

In the coming years, when people asked how Mae had died, the answer, if it didn't involve shrugged shoulders and blank faces, usually referenced her first diagnosis of pleurisy. Something had made it hard for her to breathe, her family explained. She died because, for nearly two decades, she couldn't really breathe. It was a true, if vague, explanation.

Over seventeen years, Mae endured debilitating bouts of breathlessness and exhaustion, not to mention painful and disfiguring swelling. But she did continue to live for those seventeen years. She got to meet eight of her twelve grandchildren. There is no shortage of black-and-white photographs showing her lovingly holding Billy, her oldest grandchild. She was able to watch him grow into a little boy. But she never once imagined that the flaws in her own body would become the tragic legacy that would bind them together.

EIGHT

The story of my great-grandmother Mae's illness proved to be of value to Dr. Kricket because the swelling and bouts of lymphatic buildup mirrored the symptoms of my father. But beyond those, any similarities between Mae and her grandson were superficial at best. The similarities between Mae's son Nathan and her grandson, however . . . now, there was a compelling comparison.

Nathan was, for all intents and purposes, a healthy twenty-eight-year-old in 1955 when Mae died. He was already engaged to Aunt Joanie. The family liked Joan because she brought out a little levity and rebellion in their serious brother. Once, Joan got Nathan to smoke a cigarette, which particularly impressed my grandmother. Their wedding took place weeks after Mae's funeral; they agreed it would be bad luck to cancel.

A year passed, during which Joan gave birth to a son, Michael. In early March of 1958, he was two and Joan was about to give birth to a second child when Nathan scheduled an appointment with an ophthalmologist for a little itching in his right eye and a sensitivity to light. The ophthalmologist noted a dilated pupil and suggested

Nathan look at photos of himself in order to determine how recently it had occurred. Once they determined that the dilation was recent, Nathan visited his family physician, who heard wheezing when he listened to Nathan's chest with a stethoscope. Nathan simply hadn't noticed it, and to the doctor's mind, checking Nathan's lungs had been nothing more than a routine part of a visit for an itchy eye. The doctor scheduled an X-ray that revealed the wheezing to be a more critical problem. Nathan had pleural effusions, fluid in his lungs, and this didn't seem to have anything to do with his dilated pupil. On March 28, 1958, he was admitted to Montefiore Hospital in the Bronx. Across town, Joan was alone, giving birth to their daughter.

Nathan was first tapped during that hospital stay, but while he was there, another strange thing happened. His pupil continued to dilate. His eyelid started to droop to protect itself from the light, and then the eyeball itself rotated upward. In the same way that my father's medical notes consistently referred back to his childhood heart-valve anomaly, henceforth every medical note would begin by commenting on Nathan's eyelid, which permanently sealed shut, his eyeball bulging beneath the taut skin. To a large degree, the eyeball mystery became the primary differentiator between his mother's and his own physical breakdown.

The fluid that was tapped out of Nathan's lungs in the early stages of his illness yielded no diagnostic answers. The doctors referred to it as "chylous fluid of no clear etiology"—just as they had referred to his mother's condition. However, after early testing in November of 1958, one doctor suggested that the fluid present in Nathan might not be chyle, an opinion that was later supported by the fact that there was no obstruction of his thoracic duct, where the lymphatic system meets the vascular system. No other doctor could suggest what else might have caused the fluid buildup. It was chyle, leaking for no reason. Nathan's primary organs were healthy, there was no discernible cancer, yet his body was growing increasingly bloated with chyle. Unlike his mother, whose complications emerged slowly over decades, Nathan, in less than a year, was finding his trousers tightening under the strain of visible abdominal swelling.

Nathan went home and met the baby girl they named Phyllis, but the battle for his life was just beginning.

There still wasn't enough information to prove that Nathan had a genetic illness. At the time, researchers and doctors didn't often discuss genes outside of academic and clinical environments. Still, Nathan's condition seemed obviously, if distantly, related to his mother's. So when he first became ill, his wife, Joan, made sure to dig up as many of Mae's medical files as she could to share with the multiple professionals seeing her husband. Joan and Nathan wanted answers. If they couldn't get those, they at least wanted treatments.

There are almost always treatments for monogenic illnesses, or illnesses brought about by the flawed information in a single gene. Although these treatments remain largely ineffective in the long (and sometimes even short) term, they are typically administered to desperate patients. Consider for a moment that you had to fight with your body to pump your blood correctly or produce a different enzyme than the one it thinks it is supposed to produce. What if every single cell in your body holds this "incorrect" information? It's like an already-built building that requires the replacement of every single brick, or else the whole building will fall. Where do you begin?

In 1971, a little girl was born to a young couple in Westport, Connecticut. Twelve years later, three years after that little girl lost her life to a grisly genetic condition called cystic fibrosis (CF), her father, a successful sports writer named Frank Deford, published a brief but beautiful biography of the girl in *Alex: The Life of a Child*. When I was thirteen, I read the book, having checked it out of the library because I was drawn to the picture of the cute little girl on its yellow cover. I related to the vibrant life chronicled in those pages. Alex would have been four years older than me had she lived. She would have been two years older than my sister.

Deford writes about how his new baby, his second child—the little girl he and his wife had hoped for—never really thrived. By her fourth month, it seemed like Alex was always sick. Eventually, her doctors recommended a sweat test. The principal diagnostic tool for cystic fibrosis, sweat tests measure a person's chloride levels. A person suffering from cystic fibrosis experiences elevated chloride in the sweat glands coupled with mild to severe respiratory illnesses.

To put it crudely and very primitively, cystic fibrosis involves the secretion of thick mucus that, in addition to a host of other brutal developments, clogs respiratory pathways and makes it difficult and sometimes impossible to breathe. In the 1970s, as is the case today, there was no way to instruct the body to stop producing this mucus. There was no miraculous treatment to clear or thin the mucus either. Doctors instructed parents* to follow the standard treatment for a chest cold, and ratchet it up. In short, the pot of boiling water stuffy-nosed people had been sticking their faces over for generations became a high-flow humidifier complete with face mask. But that was hardly the worst of it. Remember how people clap their fists against their chests to clear out a frog in the throat? That maneuver became an aggressive and often minutes-long chest clapping, or "percussion." Doctors and parents performed this kind of physical therapy multiple times a day on children as young as several months.

In the 1986 made-for-TV movie of Deford's book, there is a moment when the actor Craig T. Nelson hoists a small Gennie James into the air and turns her upside down, pounding relentlessly on her chest and back in the hope of loosening the death grip of that mucus. The actress calls out for him to stop, but Nelson paddles on.

---

* Most sufferers experience their first CF symptoms in infancy, making parents like the Defords the front line in their children's survival.

In all fairness, those thrusts and taps were all the Defords had to work with. Keeping your child alive is desperate and unpleasant work. Parents of CF babies in the 1970s were given only the most rudimentary of tools to try to slow the death of their young children. Truth be told, in 2016, the tools haven't gotten much better. But back then, pounding and humidifying were the best they could do.

Eleven years before Alex Deford was born and one state over, Uncle Nathan was about to get a treatment plan. And like the Defords and other CF families, he was desperate to follow it.

By 1960, Nathan was thirty-three years old and filling up with a fluid that was similar to the stuff that had killed his mother not so long ago. His young wife, Joan, was harried, teaching full-time to make ends meet and raising two young children while her husband was admitted to the National Institutes of Health. The NIH was established as the Marine Hospital Service in 1887 to study cholera, a bacterial infection that European scientists were just beginning to understand. By 1902, it had grown to become the chief U.S. public health agency. For many, including Nathan, the NIH was the last hope when battling a rare or undiagnosed condition.

Over ten months of 1960, Joan worked at her school and took care of her children—Michael, three, and Phyllis, one—with help from her mother, who lived near their home in the Bronx. Every Friday morning she'd kiss her children good-bye for the weekend and leave them with her mother or their aunt Norma in New Jersey. After a long day of teaching, she'd make the four-hour train ride to Washington, D.C., then travel to Bethesda to join her ailing husband.

Meanwhile, Nathan's pain was intensifying as the fluid infiltrated his body's tissue. Notes from his chart include disturbing accounts of the agony wrought by a thickening of the tissue in his penis and testicles, among other places.

Nathan's fluid came back more quickly and aggressively than Mae's had. The link between his illness and his mother's was so tenuous that, other than a few offhand notes about swelling in the legs of his sisters Norma and Shirley, there are very few mentions of a family trait. Though his sisters Shirley and Norma, his mother Mae, and his grandmother Ester Bloom all suffered from swollen legs, no doctors drew any significant links between Nathan and those family members.

When the NIH arranged for Nathan's release, he still did not have a diagnosis. The doctors' conclusion read: "Etiology of the basic disorder which manifests itself by chylous effusions in the thoracic and abdominal cavities is, after all these many months of hospitalization, still completely unknown."

Distraught, and riddled with fluid, Nathan was discharged. He was terrified to find himself growing sicker and sicker while his wife and young children stood looking on. His albumin levels, a measure of circulating protein, had plummeted to a point that was considered all but fatal. Albumin is a protein made by the liver and found in the blood. It is imperative for survival as it is primarily responsible for growth and repair. We need to eat protein so that our livers can produce albumin. Most of the protein going into Nathan's mouth was leaking into his body cavity before being digested and turned into albumin. Nathan was starving to death.

Since there was no way to stop the flood of chyle, doctors focused on alleviating Nathan's discomfort. They hadn't the slightest idea of what to do for a cure, so they worked with what they knew. At the time, the only known treatment for extreme cases of massive ascites, the medical term for a large amount of fluid in the abdominal cavity, was to tap it, as had been done to Nathan's mother before him. They cut a small hole in his abdomen and used a tube and manual pump to suck it out. After they removed as much of the fluid as possible, doctors had to figure out a way to replace the nutrients it carried. Years later, they would attempt to do so using intravenous

(IV) technology that continuously infused nutrition back into the bloodstream. No such technology existed during Nathan's illness, so instead they froze liters of the fluid they pulled out of his body and sent it home in bags with his wife. She prepared some every morning as a beverage for her husband to ingest orally. He had to drink it.

Tapping the fluid and drinking it was my family condition's version of the CF chest percussions, our version of holding our heads over bowls of boiling water. Doctors couldn't yet replace every faulty brick in the building that was Nathan's body, so instead they did what medicine has historically done and continues to do when faced with an untreatable monster: they size it up, take out a squirt gun, and stand there shooting. Or in Nathan's case, they made you drink your body's waste.

Nathan died at the age of thirty-four in the spring of 1961. His final diagnosis simply lists his symptoms: "Chylous ascites of undetermined etiology. Ptosis right upper lid of undetermined etiology."

My father's cousin Michael was Nathan's oldest and only son. My father was eleven years older than him, but they were close, in part because Michael, who grew up in the Bronx, moved to Pittsburgh, just three hours away from us in Columbus, in order to attend graduate school. Michael and his girlfriend Susan, whom he later married, regularly came to visit, and we went to visit them repeatedly over the years.

Michael, only four when his father died, recalls very little mention of his father's illness among the cousins in the years following his death. The truth was, no one had connected Nathan's illness to Mae's. Michael described his father's illness to Sue before they got married in 1982, referring to it as "an unusual unlucky kind of a thing." He had called it a "blocked thoracic duct," which, as a diagnosis, had in fact been ruled out. When Sue asked if the illness was hereditary, Michael told her that the doctors assured his mother that it wasn't.

He did wonder from time to time if he might get it, until he passed the age of thirty-four—the age his father had been when he died. After his mid-thirties, those fears diminished. He figured if he did contract a disease, it would be something he could beat if he stayed healthy enough. Perhaps the family lore that mostly suggested Nathan had been "sickly" throughout his life, beginning with an early bout of rheumatic fever, encouraged his son to believe that he could beat the illness, should it ever manifest.

Not a single person close to my dad has any recollection of him ever comparing his ailment to his uncle's until late in his illness. However, Michael claims that my dad referred to a connection before he even looked sick—perhaps sometime around 1990 and our family trip to Israel. My dad suggested to Michael that day that he believed he had "what Uncle Nathan had."

Michael recalls that this conversation took place during a business trip to Columbus while the two were chatting in my father's home study. He said my father was unwavering in his belief that he could overcome his condition with the right discipline. My father was healthy. He was at least ten years older than his uncle Nathan had been when he got sick. It might be the same thing that killed Uncle Nathan, but for my father, it would be different. And certainly not deadly.

NINE

All the drainage points for my father's frequent tappings, as well as IV ports, blood draws, and the other standard skin piercings of the chronically ill, were prone to serious infection. At this time in his medical journal, my dad wrote, "There is no doubt something needs to be done. I am deteriorating slowly but surely. I am pretty much a skeleton I am so skinny. I can hardly walk. I am so fatigued. Forget sex."

Meanwhile, I was doing my best to maintain a semblance of normalcy in my own life. During my freshman year of college, I finally had my first serious boyfriend. Although I could talk endlessly about boys, I had never been boy crazy in practice. I went to prom and had been kissed a handful of times, but I had never really had a serious boyfriend, until Lucas. Lucas, my freshman-year boyfriend, was the boyfriend I always thought I'd have. We had so much in common it was almost boring. We even looked alike. We had similar black, wavy, chin-length hair and tall lanky bodies. Another couple I had seen around school seemed similarly matched. When I mentioned to someone that I had trouble telling them apart, they replied, "Well, before I met you and Lucas, I thought the same thing about you guys!"

As my father grew progressively sicker, something strange had started to happen to me. My self-confidence, the certainty I had about my understanding of the world at large, was beginning to erode. I had a therapist who named it "launch-pad disintegration." She said it was a phenomenon that she believed happened to a lot of people in their late teens and early twenties when their parents died or divorced, or family units otherwise fell apart. There you were, just about to launch—when *crash!* The whole thing fell out from under you. The foregone conclusion that I would marry a doctor and become a doctor (or maybe a lawyer if I couldn't pass calculus), that I would marry Lucas who looked like me and shared a background, a religion—who was conveniently planning on *becoming* a doctor . . . well, that plan was beginning to fray around the edges.

I stopped going to classes. I started smoking a pack of cigarettes a day. I asked for extensions on tests and papers, or settled for mediocre grades when I forgot to get those extensions. I started smoking pot at parties. I stopped cleaning my dorm room. I started smoking pot every morning. I stopped imagining a future in which I would be a productive member of society. I joined a band.

That spring I brought Lucas to the Brigham to meet my father. He had just come out of surgery and was bleary from medication. If this had been any other time and any other place, my dad would have looked at both of us and genially exclaimed, "If I could create the perfect man to date my daughter, it would be you, Lucas!"

Instead, my eighteen-year-old boyfriend, looking over the hospital bed of his first girlfriend's dad for the very first time, got a different response. My father reached up and put a hand gently around the back of Lucas's neck. He brought his head toward him, and then he kissed his forehead.

Unfortunately, I did not find this moment touching. I was horrified. Lucas admitted later that he was more embarrassed than horrified, but the energy was similar. I'm sure I apologized for my drugged-out dad and ushered Lucas quickly from the room. By the next school year, we had broken up. Lucas almost immediately started dating a beautiful freshman who looked a lot less like him than I did and whom he would later marry. He also became a neurologist. Of course.

Trying to imagine what my life would have been like if I had remained the girl I was when I started college is a futile exercise. I was not that girl. I immediately started dating musicians and boys who dropped acid at Dead shows.

One of the first people I dated after Lucas was a guitarist named Ryan. I was quietly obsessed with him, and our romance lasted about a month. He called me "kookaburra," which is a bird referred to in a famous Australian folk song. It is not a compliment: "Kookaburra sits in the old gum tree, / Merry merry king of the bush is he. / Laugh, Kookaburra, laugh, Kookaburra, / Gay your life must be!"

I wasn't gay—in the sense of happy or otherwise. I wasn't merry. But I *was* at least borderline crazy. I went to lunch with Ryan one day and was so overcome with nerves that I didn't speak. He spoke and I made a noise, and he spoke again, more out of necessity than desire. My body shook during the entire, short meal. If I managed to swallow food, it was an accident. Ryan avoided me until that weekend, when I showed up at his band's show at a campus fraternity. I never went to fraternity parties and would never have gone if he hadn't been playing. He snuck out before I could find him after the show.

Ryan wasn't my only kookaburra experience that year. There was the artist at my sister's school who was on the receiving end of way too many telephone-call hang-ups, popular among casual stalkers in a pre-caller ID world. There was the blond pot dealer who lived next door to my friend Suzy whose doorbell I sometimes inappropriately rang in the middle of the night to the dismay of his roommates, one of whom still won't accept my friend request on Facebook. And there were others.

I was in an emotional holding pattern, overcome with two primary feelings—guilt and embarrassment. They had become my default emotions. Things that might have made me angry or sad had shifted to my constantly asking myself, *What did I do to cause this to happen?* I imagine that subconsciously it was a way to regain some kind of control in my life. My father's illness was taking a toll on all of us. He was exhausted as his body grew less and less able to absorb the nutrients of the food he was eating. That is, the food he didn't immediately vomit up from the pressure of the fluids encasing his organs.

We had all become so accustomed to the illness that it was our new normal, but its ability to keep us on our toes never failed to surprise us. Once, my mother and I checked my dad out of the hospital for a night to treat him to a nice dinner at one of Boston's trendier restaurants on Newbury Street. A friend had told me about an Italian-influenced restaurant, Sonsie's, that served food I knew he'd like. Outside on the sidewalk, my father's wheelchair got caught on a pothole. Dozens of people walked by us as my mother and I strained to push the chair out of the deep groove. My father was even trying to help, telling us to push left, then right, using his weakened hands to jiggle the wheel. Embarrassment exploded from deep within me. What I should have been feeling was anger. Rage. Outright fury at those people who acted like we were invisible as they forged past us. Not a single person offered to help.

My father was so sick. His tall, once-lean frame was enormously heavy with fatty fluid bloating him with pockets of swollen flesh. What wasn't fluid was hardened scar tissue. The enormity of my father—and the enormity of our situation—had become overwhelming. Despite having Dr. Kricket on the case, we still had absolutely no answers. And here was my father sitting in a wheelchair that was stuck on Newbury Street in a pothole.

When my mother and I finally managed to leverage the wheel back onto the sidewalk, we rolled the chair into Sonsie's, shaking off the momentary horror, a recently mastered skill. The restaurant was classy without being stuffy. The food was good and our banter became relaxed. Eating with my father, though, had become something of a complicated ordeal. In the moment when we had finally managed to experience a tiny spell of normalcy, all three of us forgot that we were anything but normal.

It didn't last long. My father coughed once, and immediately threw his hands up to his mouth. My mother and I leaped from our seats and started grabbing napkins wherever we could find them. My father often threw up after meals. The pressure surrounding his stomach made it hard for him to keep food down. The other diners were trying not to look. Even the waitstaff was trying not to look. But there was a large, very sick-looking man throwing up in their establishment. Someone was going to have to do something.

So they brought us our check.

I apologized as my father stared blankly at the tablecloth. I explained that he was sick. The waitstaff nodded, and I thanked them for nodding. My father's expression became morose.

On another occasion, at a dentist's office, his tapping site started to leak fluid. Although the bandage we used was guaranteed to withstand anything, the pressure behind the chyle in his gut soaked through it until the fluid poured freely from his body. We gathered towels to deal with the sticky mess.

I was conflicted. I was embarrassed one minute, ashamed the next. Mostly, like the rest of my family, I simply couldn't get a handle on what the hell was going on.

"What's wrong with your dad?" people asked. But we didn't have answers. We lacked a language or a guidebook, and instead grew accustomed to the frustration and humiliation of the undiagnosed. Years later I heard the gossip among his patients was that he had AIDS.

With guilt and embarrassment as my only collaborators, I was faltering. Everyone in my family agreed when I asked if I could take off the second semester of my junior year.

———

In the winter of 1995, my father installed a lift so he could get up and down the stairs of our house in Columbus. He also decided to sell his business, and found a buyer quickly. He'd cut his hours dramatically in the previous months, but he was managing to continue to work. He felt that if he could sell the clinic right away, he could stay on a little longer to keep its value high. Then, when he got better, he would commit to coming back and working there once again.

Dr. Steven Delaveris became the clinic's new owner, and also my father's new primary-care physician. Like his previous doctor, Dr. Delaveris was stumped by my father's condition, but he followed up on one new avenue of treatment. Patients suffering from some lymphatic conditions are fed intravenously. Since my father was deficient in nutrients, it was the next logical step. If he knew about his uncle Nathan's "chyle smoothies," he'd have known that the IV recourse was a technological leap forward. A permanent port was placed in his groin where the IV would attach. Twice a day he was fed through an IV drip—introducing nutrients directly into his bloodstream. He continued to eat by mouth when he was hungry.

The port was uncomfortable and easily became infected, but initial test results showed some slight improvements. One day Dr. Delaveris came to our house, and while he was checking on the port, he casually asked my father if he had thought about his end-of-life plans.

It was a question he asked all of his patients. He wasn't suggesting anything more than that everyone should have a plan.

My father didn't hear it like that. He heard Dr. Delaveris giving up on him. To look at my father, it would have been hard not to consider giving up. But my dad was going down fighting and anyone who didn't understand that was, as far as he was concerned, deadweight. He relieved Dr. Delaveris of responsibilities for his care immediately. My dad wanted to live and he needed the rest of us to stay very clear about that.

On February 10, 1996, my father had been forty-eight years old for nine days. I was living at home during what would have been the second half of my junior year of college. I had already been a biology major, a history major, and a psych major. I had done all the legwork, changing advisers and registering with each of the different departments. I guess changing majors was a bit like having control. Now, during a break from school, I was majoring in marijuana studies with a minor in avoiding my parents.

Meanwhile Dr. Larry Lynn, a pulmonologist in Columbus who just might have been Ohio's answer to television's Dr. House, found my father. Dr. Lynn worked at Doctors Hospital, where my father was both on staff and had been spending an inordinate amount of time as a patient over the last few years. Dr. Lynn didn't know my father, but he liked medical mysteries. When I spoke to him recently he told me that some weekends he'd troll the hospital looking for cases that had a lot of holes in them. That's what happened the day he found my dad. When I asked him how many of his mystery cases he had managed to solve before my father's, he sounded surprised that I'd even had to ask. "All of them," he answered.

Dr. Kricket's involvement in my father's case had been relegated to procedures that could only be accomplished in the lab. My father's diagnostics, so to speak, were not within her jurisdiction. What we needed was a guy who trolled hospitals looking for mysteries. So it was awfully convenient when that guy found us.

Initially, Dr. Lynn had coordinated with Dr. Delaveris to get my father on a liquid-diet regimen. Since Dr. Delaveris had been dismissed, Dr. Lynn took over my father's primary care. He had an idea. In addition to the IV nutrients, Dr. Lynn wanted to remove all the fat from my father's diet to see if he could elevate his albumin levels. Most of the protein my father was eating was leaking into his body cavity in the form of chylous fluid.

Some lymphatic conditions have been successfully managed by cutting fats completely. Fat thickens lymphatic fluid, so that the structures of the lymph system "drop it" more readily into the body cavity instead of channeling it into the bloodstream. A low-fat diet—or a diet free of triglycerides—would also prevent chyle from forming. If you are coughing up chyle, or suffocating in it, a fat-free diet can keep your body from forming that sticky milky lymph in the first place, or at least from making so much of it. Fluid will still be leaking, but it will be much thinner, more like the liquid you see when you pop a blister. It's largely more comfortable, and fundamentally easier to manage. Our main hope was that my dad's body would have an easier time channeling the protein-rich fluid into his bloodstream. A few years earlier, my father thought my mother's ice cream habit might kill her. Now it looked like foods like ice cream could kill him.

He was devastated by Dr. Lynn's fat-free diet plan, but motivated to try anything. Very soon his diet consisted of several fat-, sodium-, and taste-free soups for breakfast, lunch, and dinner, and fat-free dressings on raw and boiled vegetables; it wasn't tasty but his albumin levels were shooting up. That wasn't the only trick that Dr. Lynn had up his sleeve. Recently, things had been looking pretty grim for my father. Now his color was better. He was breathing easier, and Dr. Lynn was leading a charge in a whole different direction.

He actually introduced us to a fairly new technology, something he was calling "the Internet." By 1996, e-mail was slowly becoming a part of the mainstream. Everyone at my university had access to it on old boxy computers in the library with a flashing horizontal cursor. But at the time, the Internet was still slow, typically loading pages for minutes at a time, and hadn't quite spread into daily life. I remember overhearing someone suggesting that one day we would all be ordering everything on our computer, and finding the notion truly confounding. Why would anyone look online for an outfit when it took two minutes for a shirt just to appear on your screen?

Dr. Lynn used the Internet on his mission to research the origins of my father's illness. He was determined to find a family that had it.

And then he found them. They were in Turkey. The family was a small group of Turkish Jews who had a condition that was an off-shoot of a rare disease called Behçet's syndrome or Behçet's disease. Typically, Behçet's patients have sores all over their bodies. But this family had lymphedema! Both diseases were caused by inflammation in the blood vessels. The best part of all of this was that there was a treatment for Behçet's, and it was simple. And it worked.

Steroids.

I sometimes think of this period as one of those times that I wish could stand alone without anything coming before it or after it. On the day of the Behçet's diagnosis, it was springtime 1996, and the world was literally and emotionally on the thaw. We hugged and kissed each other and hugged and kissed Dr. Lynn. I drove to the park and sat by myself on a perfect spring day and wrote in my journal, euphoric.

"Today is the day we prayed would come and it finally has!" I scrawled across the page.

A sudden jolt in the universe had just set everything right.

Nearly ecstatic, my dad's best friend, Jimmy, went to the library and immediately looked up Behçet's disease. Hilary; Jimmy's wife, Jackie; and my mom started talking about throwing a party. Someone tossed out the possibility of combining Hilary's upcoming college graduation party with a celebration of my father's health. We began suggesting dates to take our abandoned family trip to Key West. My dad, although feeling a little better on his fat-free diet, finally got to eat some real fat.

Dr. Lynn started him on steroids immediately. He told us that the Behçet's diagnosis would be validated within the week based on how my father's body reacted to the steroids.

The next day my mother called me. She was hysterical. "You have to get to the hospital!" she shouted. "It's your father. He wants a DNR."

A DNR is a directive to hospital staff that stands for "Do Not Resuscitate" should anything go wrong. It was one of many ICU terms I had grown uncomfortably familiar with. My mother was overwhelmed. She was in the parking lot crying, getting ready to leave. So I left work and drove to the hospital. My father was staring at the wall in his ICU room.

"Daddy?" I said.

He looked at me. His face was drawn and dark.

"I heard you asked for a DNR." This wasn't like him. My dad would practically kick people out of the room for asking if he was finished reading a magazine, much less suggesting a DNR.

I sat on his bed and took his hand. "Do you remember that you just started taking steroids?" I asked gently.

He nodded.

"Do you remember that Dr. Lynn said they might make you depressed?"

He nodded again.

"Why don't you get forget about the DNR just until we see what happens with the steroids."

His eyes filled with tears. "You think I should wait?" he asked cautiously.

"Yes, Dad. I think you should wait. Let's just wait."

He was crying now and I was hugging him. "We just got this great news," I went on. "We can be happy for a little while!"

"We can?" he asked.

"Yes," I answered.

I was so sure. I was so happy; even as he cried, I was so so happy. We canceled the DNR.

That night, by pure coincidence, my father would have died when his heart stopped, forcing him into cardiac arrest. But he was saved because we had canceled the DNR.

Two days later, Dr. Lynn told us the steroids weren't working. He'd been wrong. My dad would have to go back on the fat-free diet. We still didn't know what he was ill with, but we now knew that it wasn't Behçet's.

That summer, my ex-boyfriend Ryan's band was playing a show in Columbus. My obsession had passed and I decided to go simply to get out of the house. After the show, Ryan didn't hide from me, and we went for a drive. I told him what had been going on with my family during our brief but intense few weeks of dating several months earlier. Ryan listened compassionately. He'd had no idea. It solidified a new, nonromantic friendship we've continued to this day. That night, Ryan came over and met my father. They talked for a while. My dad was a musician himself. He admired Ryan's ambition.

As I was driving Ryan back to the hotel where he and his band were staying, he turned to me and said, smiling, "I feel like I understand now why you went kookaburra."

I laughed. And then I shrugged. Because . . . well, yeah.

TEN

Since the Behçet's syndrome diagnosis had been proven wrong, no new ideas had been forthcoming. The Behçet's expert Dr. Lynn reached out to wasn't interested in exploring possible links between his expertise and what my father was suffering from. It wasn't certain that the two could, in fact, be linked, but the expert's complete lack of interest was disheartening. My father's indefatigable hopefulness was starting to show signs of fatigue. With each new roadblock, we were driven back to square one.

One day I arrived at the hospital to find my father sitting up, a glassy look in his eyes. When I entered the room, he looked at me, and for a moment, I wasn't sure he knew who I was. Then his features relaxed into a smile and a look of relief crossed his face.

"Jos, I had this dream . . ." He looked down and whispered to himself, "Was it a dream?" Then he looked back at me and went on: "You were a baby and I dropped you on your head. And you lost the ability to say no."

My dad had a lot of vivid dreams in those days, which he'd recount, but that one resonated because his lack of certainty that the dream hadn't actually been real was jarring. That, and the fact that I was getting really good at never saying no.

———————

While Dr. Kricket searched for a genetic link, my dad was getting sicker by the day. What was more, he despised the idea that he would have to spend what was left of his life eating Dr. Lynn's diet of tasteless 100 percent fat-free foods. Dr. Lynn tried reaching out to a company he had read about that was developing a food additive called "olestra" that rendered fats impossible to digest. He wanted a compassionate release of the product for my dad before it had FDA approval.

Still, as my father waited for the magic additive, wheelchair bound, physically disfigured, and deeply unwell, there was little that he felt excited about. Each new medical puzzle was beginning to feel like another nail in his impending coffin.

In the summer of 1996, a team of specialists Dr. Lynn had contacted at the University of Arizona—coincidentally my father's alma mater—invited my father to their office. Dr. Charles Witte and his wife, Dr. Marlys Witte, were the go-to team in the United States for anyone who found themselves with lymphatic or thoracic-duct anomalies. According to a U of A tribute following Dr. Witte's death in 2003, he was responsible for designing and helping to simplify "methods to evaluate and treat congenital and obstructive lymphedema."

Despite the fat-free diet and biweekly tappings, the fluid reaccumulated in my dad's body more aggressively than ever. So my parents took off and set up camp in Arizona for a monthlong stay during the early days of summer.

My mother remembers daily walks to the hospital from her hotel in the sweltering Arizona heat. She was alone again, as she had been for much of my father's illness, spending her days waiting for him to come out of surgeries and procedures. She ate alone and slept alone. Now, because of the heat, she was mostly stuck indoors. She helped the hospital staff with bathing and caring for my father. And after that, she watched a lot of TV.

Few people in the 1990s, and frankly even today, know much about the lymphatic system. Dr. Witte told me that she and her husband initially started studying it for that very reason. Their research in the 1960s showed that liver conditions like cirrhosis seemed to produce a lot of associative lymphatic fluid, more than what the body could properly circulate. No one knew much else about what happened when the lymphatic system goes awry. The Wittes' main objective was to develop or seek out treatments for the buildup of lymph, which is exactly what my father needed.

My dad's chart showed that one of the first surgeries he'd undergone since his illness in 1994 had been to counter something called a "total cavernous transformation of the portal vein," or a TIPS procedure. My father's portal vein, a main pathway through the liver, was blocked, or more specifically, it had shrunk. Doctors used a shunt, a small piece of metal mesh, to try to prop the vein open, but the portal vein didn't stay open. No one understood why.

When the Wittes met my dad, they knew about the problem with his portal vein and that something was keeping his lymphatic fluid from staying inside his lymphatic system. My father's liver was otherwise healthy, though "healthy" as a term no longer applied to my dad. Everything was growing steadily less healthy because he was starving, as every day vital nutrients leaked out in fluid that never made it to his cells.

The lymphatic vessels are difficult to see. They are thin and colorless, unlike the colorful and more substantial veins and arteries. My father underwent a simple test called a "lymphoscintigram," in which doctors inject a radioactive element and a bit of blue dye between the toes. The strong pressures in the lymphatic channels should immediately pull the dye up and through all the vessels of the system. A camera records the process. In a healthy system, it should take only about twenty minutes for the dye to run clear through the body. If there are any leaks, however, the camera can pinpoint exactly where some of the fluid exits the system. It can even trace where the fluid goes: if it lingers in various tissues, or pools in the ankles.

My father's test didn't get very far. It lasted less than a second. His lymphatic system was so weak that nothing moved much beyond his ankles and most of the fluid just settled near the injection site in his foot. The vessels in our bodies rely on pressure. When that goes, everything else goes too. There just wasn't enough pressure in my father's vessels to channel much of anything.

The Drs. Witte were at a loss.

Every test they administered came back inconclusive. Dr. Charles Witte's opinion lined up with earlier opinions my father had already heard. Dr. Witte believed that tapping the fluid would grow increasingly difficult as scar tissue built up and trapped it. Soon tapping would become both impossible and "debilitating." To manage this condition, Dr. Witte thought there were two options. The first was to reduce lymph formation, and the second was to accelerate lymph return.

There was a small team of doctors in Italy at the time testing a new procedure to cut and attach a lymphatic channel directly to a vein. Lymphatics are heavily "fenestrated," meaning they contain a lot of openings. This is why it's so dangerous for cancer to enter the lymph nodes. The lymphatics are full of holes, and once cancer gets into them, it has a lot of places it can get out from again.

Today, doctors sometimes use glue to fill in holes in the lymphatics where there are obvious leaks. These treatments have been tested with limited success on infants born with congenital lymphatic defects, or who suffered heart conditions in utero and are born with lymphatic swelling. The goal of the procedure is to encourage their bodies to form collateral pathways—alternative channels—around the glue and to continue to move the lymph through healthy parts of the system. In 2016, long-term studies still weren't conclusive. So in 1997, all of this was largely fantasy.

But my father wasn't asking for a conventional cure for his unconventional illness. He was okay with long shots. In his case, if all this leaking fluid was channeled directly into the veins, it was possible the body would find a way to reabsorb it, and keep it from leaking. Everyone was on board with trying to get my dad into the Italian study, but there remained the question of whether or not he would be strong enough to make the journey to Europe, much less survive the procedure itself. In the end, he was never invited to Italy, and no such procedure was offered in Tucson.

However, the doctors did cave after my father spent weeks begging them for proof that his miserable fat-free diet was actually helping. Dr. Witte sent him to a nutritionist who decided the fat-free diet simply wasn't doing enough to justify my dad's discomfort. It's possible the nutritionist just felt compassion and didn't see a good outcome for my dad.

But back in Columbus, Dr. Lynn insisted he had statistical and medical proof that the fat-free diet was elevating my dad's numbers. This indicated that at least some of the nutrients that had been disappearing into my dad's lymphedema were finally breaking down and getting into his system. This diet showed at least a tiny bit of possibility that my father's starvation could be turned around or at least slowed down.

My dad was a man who desperately wanted to live, and his hatred of the fat-free diet had begun to cloud that objective. When Dr. Witte's nutritionist suggested that the diet was probably ineffective, my starving and miserable father took it as good news and ran with it to the nearest purveyor of the Chicken McNugget. When the compassionate release of olestra came through, it no longer mattered.

By the time my parents returned to Columbus, my dad had been eating every fatty food he could get his hands on. His numbers looked terrible to Dr. Lynn.

That spring, although I had technically taken a semester off to help out with my father, I got a full-time job at a drug and alcohol day-treatment center that was run by Jackie and Jimmy. It was ironic for me to work in a rehab, as I was enjoying a thriving pot habit and the occasional opiate-based painkiller that I stole from my father. My social life also stayed active. I often went down to Cincinnati to visit my best friends, Amy and Jason, who were going to school there, or they came to Columbus to stay with me during many of the weeks my parents went away.

My parents came home from Tucson in early June 1996. By then, my father could barely walk. He used his chairlift to take him downstairs and upstairs, but he could no longer make it up and down the steps in the basement, where his beloved music and home-theater system now languished. He was becoming increasingly depressed. Jimmy and I drove to Best Buy and bought him a small stereo to listen to in his upstairs office.

When he dies, I caught myself thinking, *I'll take this little stereo to school with me.*

One day after my parents' less-than-triumphant return from Arizona, my father announced that he wanted the four of us to go for a walk. It was a perfect summer afternoon. The air was warm but not hot. The low humidity was some kind of Ohio miracle. We helped my dad into the chairlift, and Hilary carried down his wheelchair. Outside the kitchen door, she wheeled him down the newly installed ramp in the garage and down the driveway.

We walked up our street, away from the house my parents had bought twenty years earlier, in 1976. My parents told us the story of driving around the strange new neighborhoods of Columbus's east side when my mother was pregnant with me. They had fallen in love with the red brick of this house, and its welcoming white porch. They especially loved the neighbors, almost all of whom we remembered by name. We walked by the Mirvis house, where the Shkolnik family used to live. Hilary and I would knock shyly on the door to see if Todd and Josh could come out and play tag on summer nights, along with Maxwell two houses up and Michael and Sean across the street. We strode by the Olivers', where Mrs. Oliver used to let us pick a flower from her spring garden. We walked past the Spatts', whose youngest daughter, Jodi, used to babysit us, and up to Jimmy and Jackie's house, where we still spent a lot of time. We stopped at the top of the block and I took a picture of my family: my dad in his wheelchair with Hilary's hand on his shoulder, my mom beside them. Everyone is smiling.

That night, my father couldn't breathe. My mother called 911. Emergency trips to the hospital by ambulance had become pretty much routine for us. The drivers asked the same questions they always did. Whoever rode with my dad that day provided four years' worth of flimsy answers to the EMT: No, he isn't geriatric. He's forty-nine. Yes, his heart is strong. No, he doesn't have a diagnosis. Yes, you should give him oxygen. No, he isn't mentally impaired. Yes, I said forty-nine. No, he can't answer your questions, because as you see, he can't breathe. Yes, that's right. No diagnosis. Then there was nothing more to say that my father's racking gasps for air didn't articulate better.

ELEVEN

By August, my father needed to be intubated. He had all but moved into the ICU. I, on the other hand, was serving ice cream. My best friend, Amy, got me a seventeen-day job at the Ohio State Fair, less than a mile away from the hospital. We worked in the dairy building. I woke up and drove myself to the crowded fairgrounds wearing a cow-print baseball cap, oversized white "Dairy Barn" T-shirt, and whatever shorts were the least covered in ice cream from the day before. There, I spent the day serving dairy products to warehouse-sized crowds.

The rules of the dairy building were simple. Show up every day for the whole length of the fair day—nine A.M. to ten P.M. Wear the funny cow-print uniform. Sell the food.

Amy's brother, our manager, agreed that I could leave the job and come back when I needed to visit my father in the hospital. There, I usually sat beside my father's still body and watched TV. Sometimes, if he woke up, we tried to communicate. He opened his eyes wide, momentarily battled the machinery forcing air directly into his lungs, and reached for me. I rubbed his head and reminded him of where he was.

I never wanted to leave the fair. My job there, despite the horrors awaiting me at a nearby hospital, or perhaps because of the juxtaposition of them, was truly one of the greatest experiences I've ever had. I was working alongside my friends, Amy and my other best friend, Jason, who was also working there. We were joyously together, surrounded by fair rides, games, and junk food. We wore our ridiculous work outfits, chose the music to blast, and flirted with an enormous thirty-to-forty-person staff, all of them our age, mostly lesbian members of another manager's baseball team—although none of us was specifically a lesbian. Amy, though, did nurture a flirtation with a very cute skater named Erica. Jason, who is gay, started seeing one of the only other men on staff who was too.

And I met Jeromy.

Out of the gate, I couldn't get a handle on the crush that took the wind out of me and explained, in stark relief, why we call these infatuations "crushes." It made kookaburra look downright tranquil.

Jeromy, like me, didn't follow the all-day rule of the dairy building. He'd been working there for a long time, and had some leverage. By day two of the fair, I sat out back on the milk crates and waited for him to arrive, smoking compulsively until he did. I was a walking lovestruck cliché, unable to eat, overwhelmed by the blood that filled my face whenever he spoke to me. The only thing I had going for me was that the feeling seemed to be mutual.

Our conversations mostly sounded like grunting. "Cool," I'd say in my cow-print baseball cap. "Yeah," he'd agree, clearing his throat.

We smoked and we talked poetry, which was as fine a basis for love as anything else during those seventeen days. Jeromy was the first person I had ever met who had dropped out of college and was proud of it. What's more, he had done it after nearly four years of attending. He was smart. To a confused, twenty-one-year-old English major, he was borderline genius.

At some moments, I believed he was Jack Kerouac reincarnated. He was a skeptic, and he pulsated with rebellion. With his white

T-shirts, cigarettes, and short dirty-blond hair, he looked a little bit like James Dean. His blue eyes never seemed very far from tears. He was quiet and hyperfocused. It felt good to get caught in his gaze.

To be fair, if Charles Manson had walked into the dairy building that day instead of Jeromy, I might just as quickly have asked to sign up with the Manson family. I was ripe for a cult—any new system of belief I could get my hands on. Jeromy grabbed hold of my flimsy belief system and began to reshape it with his monologues about "society" as a construction rather than an immutable fact, and how "Do you believe in God?" is as dumb a question as "Do you believe in the conversation you are having about believing in God?" Because there it is, God and the conversation, whether or not you *believe* in them. I found myself stricken with awe every time he spoke.

One day I offered him a ride as he was leaving. It turned out he had another job in the kitchen of a nearby bar and grill. On the drive over, I told him that my dad was dying. He just sat there, reassuringly unfazed. Death didn't shock Jeromy. Or scare him. Unlike everyone else my age, who suddenly lost the ability to make eye contact with me when I said my dad all but lived in the ICU, Jeromy's reaction was unconventional. When I told him how hard we'd been fighting for my father to live, he responded simply, "Why?"

Death, this thing that had all of us running so scared, Jeromy found inspiring. To him, the reality of death posed a possibility. It was a reaffirmation that nothing mattered, that in the end, we were all worm food anyway. He was nonchalant about it, so much so that he didn't seem cynical—he seemed matter-of-fact.

———

By the last day of the fair, two weeks since we'd met, Jeromy and I were on the verge of becoming a couple. He had broken up with his girlfriend of two years, and I had professed my love.

After the end-of-the-fair party, he accompanied me to the hospital, where we stood together over my father's body, asleep and unmoving in the eternal twilight of the ICU. I looked over at Jeromy's sincere gaze as he looked at my father for his first and only time. His expression wasn't pitying or callous. He didn't say a word to me or to my dad.

Jeromy was a kind of antidote to my father's dying a meaningless death; in my dad's own view, he was dying a loser, a miserable casualty of the battle he'd fought so long and hard to win. I needed to believe that his death, if that's what was coming, could mean something important or even wondrous. If Jeromy had been peddling heaven or reincarnation, those might have been just as good. Because I needed to believe in something positive and Jeromy gave it to me by pointing out that death, if nothing else, meant peace after the pain of dying—or maybe he meant the pain of living—in my father's case, either now applied. As a machine breathed for my father's struggling lungs, I wanted that peace to come for him as soon as possible.

Suddenly I had a narrative so much better than the one that ended with devastation and loss. I could no longer fathom the importance of my father's staying alive, not like this, like one already dead. Jeromy's skin in the blue light of these machines looked like porcelain. I was so in love with him right then that if the ICU had been a set in a musical, the frozen bodies that filled it would have burst into song.

Then I looked again. There stood Jeromy—the antithesis of my father and everything he'd ever believed—a rebellious college dropout, with a bleak near-disdainful view of life. Jeromy looked up at me then and stared solemnly into my eyes. Spending a moment with a dying man and his daughter was spiritual for him. I tried to smile, but I was overwhelmed by a single thought:

Holy shit, Dad, I'm so glad you're not awake for this. You would seriously—seriously—*hate this guy . . .*

TWELVE

Even though I went to a liberal arts college in the nineties, I never put much stock in Buddhism or the practice of Zen. At twenty-two and in the middle of falling in love, I was hooked on wild euphoria only slightly less intense than deep despair set to a Sinéad O'Connor sound track. Thanks to my relationship with Jeromy, I was chasing the rabbit just to see how high I could get, and I didn't need drugs to do it. The idea of tempering my emotions, seeking any kind of balance at all, just seemed incomprehensible.

On the other hand, one Zen idea I did embrace was this: fear is a complete waste of time. I was learning at a breathtaking rate that everything, even the worst things, is always worse in the anticipation than in the reality. Once the thing happened, it was never as bad as that blinding, paralytic fear that preceded it. I had learned that when the worst was happening, fear no longer really existed. I learned it because every day, it seemed like the worst was happening.

You might be thinking, *That's bullshit. When you are under attack, you're afraid.* But even this fear is about the existing attack intensifying, or the possibility that you might die. It isn't about fearing the attack that is already under way. When things reach their absolute worst, something wondrous occurs: a bearable emotion like sorrow or resignation settles over you. I'm not talking about afterward, when the horrible thing is in the rearview. I mean right then, during the unfolding of the horrible thing, you might even get lucky and go numb. Occasionally, you might get a burst of power or fight; that's when the anger takes over and fear is gone.

The worst, it turns out, is always the thing you feel before *the worst* happens.

There was so much fear surrounding those last months that my father spent in the hospital. My grandparents no longer came to visit. They now only came to Ohio if my mother was unable to care for my father for a period of time and needed them to tag in.

To my mind, my grandfather had always been a physically powerful man. He enjoyed spending time with his grandchildren. I remember his full but quiet smoker's laugh, a light rattle from deep in his chest. I recall him removing his bridge to reveal large potholes in his teeth, to amuse us on a whim. He was as big as his giant La-Z-Boy, where he sat and read to us. He smelled like his pipe collection. He wore the same pair of black Converse sneakers around the house until his big toe came out of one of the holes. He brazenly displayed a framed poster of Victory cigarettes from the 1960s, with a drawing of a naked blonde posing across it, above the desk where he worked throughout his retirement. We all fought over that poster after he died.

As I walked past my father's room in the hospital, I watched my grandfather shave his dying son. He gently smoothed on some shaving cream. Carefully, he raised his son's chin and scraped the razor over several days' worth of stubble. When my sister and I arrived at the door, we saw our father crying.

"I'm scared," he said softly to his father. "I don't want to die."

Hilary and I stood at the doorway and watched as our grandfather's hands shook ever so slightly while he continued to shave our father's face, tears cutting pathways through the white foam.

One condition of my semester off was that I return to school the following fall, no matter what. My father had been particularly insistent on this point.

"People who take time off from school, statistically, don't go back," he'd repeatedly warned before he agreed to the hiatus nine months earlier. But by mid-September, his health was so precarious that I was fielding regular emergency phone calls from Ohio when I was back in Boston. By the third week of school, I flew back to Columbus after my father suffered a stroke. Once he stabilized, I flew back to school again.

The day after I returned from the stroke trip, I was putting on makeup for a Friday night out with friends at a pub we frequented in Davis Square. Sure enough I got another phone call. My sister told me I'd have to get another ticket and fly back home again the next day. Dr. Lynn decided it was time to put my father under palliative sedation: a compassionate coma for the dying.

I arrived at the Burren, our neighborhood pub, an hour late. My friends were already drinking pints. I slid into the booth at the back of the bar where they sat and lobbed off the following in one run-on sentence: "Sorry I'm late my dad just slipped into a coma what does a girl have to do to get a beer?"

My three friends stared at me in silence.

The next day, when I arrived in Columbus for the third time in almost as many days, the hospital had convened an ethics committee. This time my father's kidneys had failed. He had survived, but there was a question about whether or not we should keep him on dialysis. If we took him off, poisons would back up into his body. He would feel itchy, but we were assured that dying in this way wouldn't be painful.

I was on board. It was time to call it a day. My sister and mother lingered in the background. They were devastated. I felt manic in my desperation to end all of our suffering. The summer had been relentless and exhausting as my father's nutritional levels plummeted. His lymph had become impossible to remove, trapped by hardening scar tissue, leaching hydration from his body. The process was finally taking out his organs one by one.

I was angry when the ethics committee voted to keep him on dialysis. Jackie and my sister hugged me as I ranted and begged. Finally, Dr. Lynn decided to wake my father from his medically induced coma. It was time to let him weigh in on his own fate.

My sister recalls Dr. Lynn calling out loudly, "Bill? Hey, Bill! Do you want to die?! Are you ready to die?!" He repeated it over and over until my father was shaken awake.

BILL, HEY, BILL, YOU WANT TO DIE? ARE YOU READY TO DIE?

I don't know about the last time *you* were woken up from a medically induced coma, but if I was the one receiving this wake-up call, I'd probably ask to be shot directly in the head. Instead, my father blearily looked around the room, catching the exhausted and horrified faces of his family.

"What does Joselin think?" he asked, harkening back to the day when I had rushed in to stop the DNR after he had been put on steroids.

"You don't know, Daddy," I began this time. I was a new person. I no longer saw death as something horrible. Maybe life and death were too large to encompass with concepts like "bad" and "good." My father, even now, was trying to fight. So I told him emphatically, "You've been out of it all summer. I mean, it's September!" Did he even know that? Hilary was bawling. My mother couldn't look. "I think it's time, Dad," I said. "You've been through so much . . ." And so had we.

I know he didn't mean what he then said. Even though his voice was almost inaudible and his words were coming out in percussive breaths from the multiple tracheotomies that had scarred his throat, it echoed when he looked directly into my eyes and spat his last words to me: "Fuck. You."

Like I said before: my dad was going down fighting and anyone who didn't understand that was deadweight.

I didn't go back to see him that night. The next day, my mother and sister asked me to go with them to the ICU. I had made my father a mixtape of his favorite songs: "Red Rubber Ball" by the Cyrkle, "Homeward Bound" by Simon & Garfunkel. We cued up the song "Dream" by the Everly Brothers. We stood around his bed and the three of us started singing, "I need you so/that I could die/I love you so . . ."

Then something wonderful happened. We heard him trying to join us. It was weak and guttural but unmistakable: "And that is why/ whenever I want you all I have to do is dream."

I returned to school again, but only days later a seizure left my dad with what we were told would be the mental capacity of a four-year-old. That is, if he woke up again at all. I flew home.

Two days after that, my grandparents, my mother, Hilary, Jackie and Jimmy, and I were eating Chinese food at the dining room table of my parents' house when we received the call from the hospital. We had been there all day. Now the doctors said we should come back quickly. It wouldn't be long. Jimmy drove with my grandfather in the front seat, and my grandmother, Hilary, and me in the back. Jackie and my mother stayed at the house.

When we got to the hospital, everyone went inside, and I stood in the parking lot smoking a cigarette by myself. An enormous harvest moon hung low in the sky. I was watching it just as a single leafless tree exploded with blackbirds, cawing loudly as they erupted into the dusky light of the fading day.

You're already dead, I thought.

My father's final autopsy report lists his date of birth as 2–1–47. Then it says simply, "Expired: 9–25–96." Like a carton of milk.

There are multiple causes of death listed. But, as they say, everyone dies when the heart stops. My dad's autopsy describes "1000 cc. of creamy white chylous fluid" filling his abdomen. The surgeon who assisted on the autopsy later told Jimmy that my father's organs had all but fused together from the five-yearlong pressure of that lymph.

Hilary didn't want to see his body. I did. He didn't look like my father anymore. The thing that made my father, my father was no longer there. I wasn't particularly religious, but what I was seeing convinced me that there's hope that our existence doesn't fully end at our death. Our bodies end. But they are not the things that make us, us. I knew this as I looked at the strange flattening of this bloated, unfamiliar body. The nurses had placed a stuffed bear under his arm that had arrived with one of the many baskets of flowers. My father would have never slept with a bear. I removed it and put it on the table.

Two days later, his funeral was standing room only. There were people all over the place, all the way to the far back wall. The week before, Jackie and my mother had gone to pick out a casket. My mother didn't want to go into the funeral home, so Jackie had to choose. Jackie knew my dad would want the top of the line, but my mom would want the cheapest. She chose something that split the difference. She got it extra long so he wouldn't have to scrunch.

At the funeral, my uncle Norman, my grandparents' middle son, told me he didn't have the gene. At the University of Southern California, where he'd been tested, they hadn't heard the telltale heart murmur. It had probably been a great relief to him, but frankly, I hadn't even considered that he *could* have the gene. I mean, of course he could have the gene. I just hadn't considered it. He and my dad were so different, not that that made for a scientific litmus test. I was glad for him.

Four eulogies were read one after the other: Jimmy. Uncle Norman. My dad's childhood best friend.

Finally, the rabbi:

Bill Linder saw life as full of opportunities and seized them. Now, much too soon, and after too much suffering, he is gone, having left another lesson as well, about how to die with courage, surrounded by love.

We drove in a limo to the cemetery. As we pulled up, the dreary autumn sky suddenly unleashed a torrential rainfall. We were ushered underneath a tent where a burial hole was sunk in autumn's hardening ground. It was difficult to hear the rabbi over the noisy tapping of rain on the canvas. My father's extra-long casket was lowered, and per Jewish custom, each of us was handed a shovel to throw dirt onto it. As the first shovelful landed, a bright flash of lightning and a resounding thunderclap erupted nearly simultaneously. Then everything went quiet. The rain stopped and the sun came out.

THIRTEEN

After the funeral, I drank wine with the rabbi and smoked cigarettes by the side of the road out of eyeshot of my mother. Amy and Jason had driven in from Cincinnati for the funeral and left early to drive back.

Later, Jeromy showed up. We had been cultivating our relationship, seeing each other on my regular trips home from school over the four weeks since the fair had ended. We wrote love letters and scheduled long phone calls like dates. Our relationship was still new. Jeromy sat at the kitchen table, far away from the other guests. I served him a plate of food. We discussed the fact that he had a plane ticket for Boston to visit me at school leaving that very night, but clearly, I wasn't in Boston. He had contacted one of his friends who lived there and arranged to stay with him. I assured him I would try to make it back before he left.

Getting those arrangements right struck me as deeply important that evening, more important than mourning with my family, which, in all fairness I felt like I had already been doing for several years nonstop. As Jeromy and I said good-bye, I remember feeling deeply upset about the possibility that I would not see him again that week.

My father's death had occurred during the week of a holiday. This meant we couldn't observe the ritual week of mourning known as shiva (Hebrew for "seven" as in "days of mourning") because the holiday was set to begin the next evening. The celebration of the harvest holiday trumped the ritual observation of death. When the rabbi explained this to us, I'm pretty sure my reaction was an inappropriate victory whoop. I probably followed it up with something like "That's cool. I'll just head back to school, then. You know, I've missed so much . . ."

I didn't actually care about missing school. The joy of my burgeoning relationship had me glowing. My father had died and I felt almost thrilled—not that my dad was dead, but that his suffering was over, and *my* suffering was over. I think my mother's suffering, and even my sister's, were just getting started. I, though, was bolstered by the pleasure that accompanies falling in love and being twenty-one. It felt like freedom.

———

The morning after the funeral, my uncle Norman and his wife, Ellen, took it upon themselves to begin the difficult task of cleaning out my father's closets. I didn't care at the time—or even consider that one day I might want some of his stuff. My aunt and uncle didn't believe it would be good for any of us to live with the belongings of a dead man staring us in the face. My aunt had read somewhere that it kept you from moving on.

My sister looked on in horror as Aunt Ellen filled enormous garbage bags with my father's clothing, books, jewelry, and medical equipment. Our defeated mother muttered, "Sure, take whatever you want," as they gathered up items for their sons and daughter, piling up his nicest watches, ties, and jackets. They were, after all, his brother, sister-in-law, niece, and nephews. It made sense to give his things to them, didn't it? And what they didn't get, some person shopping at Goodwill would make a good home for. Years later, we have often looked back on that afternoon with deep regret, having lost nearly all the material belongings from our father's life in a matter of hours the day after his funeral.

Later that night, my mother, sister, and I sat alone in our quiet house. We huddled together in the room that we had always called "the library." It was my father's office, with large wooden bookshelves lining every wall. Our dad, my mother's husband, was gone. We held hands. We hugged and cried alone, together. For the first time in my life, we were a family of three.

———————

I was happy to leave the following morning and spend most of that week in Boston with Jeromy. Our relationship was less than two months old, but it had been particularly eventful. We used our early days of falling in love to discuss and explore every horrible thing that had ever happened to each of us, a habit we never outgrew. Jeromy's revelations included an insidious history of abuse that had funneled down the generations of his family tree. He told me that one of his cousins had tried to commit suicide. He had taken sleeping pills, but the dosage wasn't enough to kill him. Jeromy told me that next time he had said he would do it two ways, like take the pills and shoot himself, just to make sure.

Jeromy was the coolest person I'd ever met. He had fronted a band in high school called the Deceitful Peaches. I didn't mind that he had baggage. I figured I was up for the challenge anyway, after the last few years. I didn't even bat an eye when he confessed that he'd been trying out heroin to see if it could help him produce better art. In my effort to seem the perfect codependent, I probably just nodded and agreed that it was best to try out every option. After all, who was I to judge?

My newest philosophy was: *we are beholden to nothing!* I understood that the only thing I wanted to do for the rest of my life was explore, find greatness, and die fat and smart. I confess I was happy, even verging on euphoric.

Jeromy moved into my Somerville, Massachusetts, dorm room that
October. He got a job at a nearby restaurant, cooking in the back.
I took a lot of poetry courses. The fact that he worked nights made
it easy to keep our relationship separate from my college social life.
Jeromy couldn't stand frivolousness and everyone I knew was in college and around twenty-one. The two worlds could only mesh when
everyone was on psychedelics. Mostly I spent time in one world
(school) or the other (Jeromyland).

Over winter break, my mother, sister, and I took that trip to Key
West we had been planning when my dad first realized he was sick.
Jeromy stayed with his parents in Columbus. I had become uncharac-
teristically brave since my father's death and my relationship with Jeromy had begun. Jeromy suggested that if I was going to go someplace
as lame as Florida, I should make sure I found the art, which to him
meant the *real* places with the *real* people, and also Hemingway. I had
grown up in a middle-class suburban world. I was not a risk-taker. In
Key West, though, I was keen to leave the manufactured spectacle of
Duval Street. With my notebook and trusty camera, I strolled through
neighborhoods where cocks fought mid-street. I was surprised that I
didn't feel self-conscious among the locals in my touristy dress and
purple off-brand high-tops. I sat in public gardens earnestly reading
Milan Kundera and Anaïs Nin. Although it was different from any
trip I'd ever taken, it wasn't fun. Not even a little bit.

Key West was one of my dad's favorite places. He and my mother
had loved spending time there over the years. Now the three of us
walked around bereft, even as we tried to blend in with the tie-
dyed T-shirts and ubiquitous plastic parrots. The thing that we just
couldn't get past, the thing that kept showing up three meals a day,
five-star restaurant or Margaritaville, was the goddamned empty
chair at the table. There is no such thing as a three-top.

FOURTEEN

In the spring of 1997, as my graduation neared, I decided to take advantage of my college health plan and go in for one last checkup before adulthood. I peed in a cup, had some blood drawn, and left. A few days later, I received a call asking me to come in to speak with a doctor. I wasn't even remotely concerned, although it's never a good thing when a doctor wants to be able to look you in the eye.

After a moment of flipping through my chart, the white-haired man looked at me and said, "Your platelets are alarmingly low."

Platelets are one of four things that make up blood—along with red blood cells, white blood cells, and plasma. The primary function of platelets is to clot your blood after you get cut and oxygen hits it. Platelets change shape, become sticky, and form a clot to seal the wound.

"How low?" I asked.

"Well, very low," he answered. "If you got into a car accident, I'm afraid you would bleed out."

I must have blinked blankly several times. My father had been dead for less than six months after a four-year illness. I wasn't sure I could handle what was coming.

"I feel okay," I told him.

"We'd like to run a few more tests," he told me.

I returned a week later to learn that nothing seemed to be amiss, and was given a prescription for iron pills.

I called Dr. Kricket to let her know what was going on. She comforted me by saying, "Look, kid, your blood clots, right?"

"Right," I agreed. I mean, so far it had never *not* clotted.

"You're anemic," she added. "Eat more leafy greens."

My sister, Hilary, meanwhile, was back living in Columbus, where she was feeling exhausted. Most doctors would have thought she was depressed, lacking direction, and mourning the recent death of her father. But not Dr. Steiner.

Dr. Carla Steiner was our family doctor. She was one of my father's close friends and colleagues. My dad had always insisted he couldn't be objective as a doctor for his own family. Over the years, my mother, sister, and I have joked that perhaps Dr. Steiner had a crush on him, which is why she always seemed so eager to please him, but was so remarkably unkind to the rest of us. In truth, though, Dr. Steiner was happily married. Our father had played guitar and sung the first song she and her husband danced to at their wedding. Her apparent unkindness was, more likely, due to the fact that she just wasn't a girl's girl.

Hilary decided to get a routine checkup. Dr. Steiner drew blood and, much like the doctor at my university health service, discovered her platelet count was alarmingly low. However, rather than call her in and look her in the eyes, Dr. Steiner decided to call our mother and tell her that it was within the realm of possibility that Hilary had leukemia.

Of course, my mother heard, "Hilary has leukemia."

My mom called everyone. She contemplated reinstalling the chairlift. Maybe we would just be that family who contended with dire illness after dire illness like a soap opera without rippling muscles and perfect lip liner.

But Hilary didn't have leukemia. Dr. Steiner downgraded her diagnosis to mono. Then finally landed on anemia. Like me, it sounded like Hilary needed to eat some spinach. That was all.

In May of 1997, I marched with my university graduating class, which is different from actually graduating with them. I had to finish one more semester before I could officially graduate. My grandparents flew in for the event. Jeromy and my best friend, Amy, came as well. We all pretended like it was a real graduation.

After the ceremony, my family gathered in Columbus for the unveiling of my father's headstone, another Jewish custom, which usually happens a year after a burial. In order to coordinate with my Phoenix-based grandparents, we moved the date up by six months.

At the unveiling, my sister burst into tears and ran dramatically away from the ceremony. I watched her as though we were in a movie. It seemed like something people did in fiction, not in real life. That's how emotionally distant I was from my sister's grief. After all, death meant peace. There was no need for drama. Jackie hurried to her side and hugged her until her heaving subsided.

The stone my mother had chosen read simply *I remember the waters*. It bore my father's name, *Dr. William Linder, D.O.*, and the dates of his birth and death, *February 1, 1947–September 25, 1996;* he was four months shy of fifty. The other half of the stone was blank, holding a spot for my mother, who, frankly, was already starting to seem emotionally buried next to him.

The quote—*I remember the waters*—was from a song my father had written that I loved. I thought my dad would like having his own words immortalized, and my mother and sister had both agreed. It was appropriately poetic. What did it mean? Was my dad "the waters"? It was corny but sweet, like a good epitaph should be. The song went:

> *I remember the waters.*
> *I remember them crystal clearly*
> *Flowing brightly by.*
> *And I remember the feeling*
> *Of a dew-set autumn morning*
> *A few years gone by.*

The lyrics, though perhaps a little vague, seemed full of feeling, ripe with nostalgia . . . or just better than *Here lies William Linder, husband, father, son, brother . . . carrier of an unfortunate gene.*

The night before I left to finish my final semester of college, Princess Diana was killed in a fiery crash. My only thought that night was for her boys, both of whom I figured would soon be battling drugs and alcohol.

As I mourned the loss of a young mother that night, Jeromy and I began another of our many fights. I didn't entirely understand his complicated relationship to life, even if I respected it. I once confronted Jeromy with a rope he had tied in a noose hidden among his stuff. He told me innocently that he used it to carry his records, and although I couldn't understand how a noose would simplify carrying his records, I found it darkly charming. At least he listened to good music. Mostly, I felt like I was always trying to catch up and keep up with Jeromy. His brilliance outpaced me, and I would never comprehend his pain. He moved back to Columbus and into my parents' lake house, which had been pretty much abandoned for the winter. I gave him one explicit instruction: he was not allowed to kill himself there. Fortunately, he accepted the condition.

Several months later, we broke up and he moved to Arizona to live with his parents.

———————

After I *really* graduated the following December, I moved home and started waitressing. I felt myself recovering some of the "Old Joselin," the one I had been before my father's illness, who raised her hand in class and completed her course work without telling the teacher she needed to be excused because she was having a personal crisis. I understood that the "New Joselin" needed health insurance. It didn't matter that I was only twenty-two and had never really been sick, unnamed genetic variant aside.

I chose a policy and my mom wrote the check. I mailed in the paperwork, but a few weeks later, I learned by letter that my insurance application had been rejected. Chalking it up to a misunderstanding, I called a number and set up an appointment with an insurance agent. As I sat across from him at his desk, my mother at my side, he typed a few numbers into an enormous desktop computer. He winked at me while the computer lazily sent along its information. Then the good salesman's face fell.

"Do you"—he looked up, suddenly awkward—"have a heart murmur?"

How did he know?

"Yes," I replied, "but it's functional."

"I'm sorry," he said, a little bit sweaty, "but the state of Ohio's preexisting-condition code precludes you from getting health insurance."

The only two people who knew about the murmurs were Dr. Kricket and Dr. Steiner. No other doctor had listened closely enough to my heart to hear it. It was very difficult to hear through a stethoscope. Dr. Kricket's study wouldn't have been made public, and certainly wouldn't have made its way into the state of Ohio's brand-new medical database. But Dr. Steiner . . . She seemed like she might be the kind of rule follower who just might put us in this position, whether or not our heart murmurs meant imminent death or long and healthy lives, simply because she felt it was her duty.

I left the agent's office that day, and remained, for the next ten years, without health insurance. We also, as a family, made plans to change doctors.

Although Jeromy and I had broken up, we maintained a tenuous relationship mostly through letters in which we'd discuss Russian literature and "real problems" like Siberian prisons and Napoleon.

Mostly he helped me flesh out plans to move to Europe with Amy. We'd decided to move to Spain after she completed her five-year acting program. I worked all spring and summer, cashed in my bat mitzvah bonds, and, with our pooled money, bought us two one-way tickets to Barcelona. I signed us up for a monthlong class to learn how to teach English that also promised to help us find jobs upon completion.

I don't believe I would have done something so out of character as move to Europe were it not for Jeromy. Our relationship status was vague, but the ocean and the hundreds of miles between us were concrete. He seemed impressed that I made the move. He'd often said he would like to spend time in Paris like Hemingway, Henry Miller, and Ray Bradbury. Europe appealed to Jeromy on a literary level, if not a literal one. But I was the one actually going there.

I have to admit I was relieved he never asked to join me. Life with Amy shone in stark contrast to life with Jeromy. While he was always deliberate and somber, Amy was spontaneous and funny. Together, we were confident and brave. Everything was surmountable with Amy along.

When Barcelona didn't work out and we couldn't find work, we took a train to Prague on an offhand suggestion from a friend who'd once had a Eurail pass. We didn't care that we didn't know anything about Prague, or anything about Europe, for that matter. When our train stopped at a station on the route to our destination, we got off. We eyed the large gold letters on the wall: MÜNCHEN. I called ahead to our contacts in Prague to let them know when we'd be arriving.

"Um, we're in *München*," I said, landing hard on the word "*munch*." "Our train arrives at eight . . ."

Our train to Prague wasn't departing for another four hours, so we set off to eat dinner in this strange, new, but exciting town. Over pizza and beer, our waiter inquired politely, "So, what do you girls think of Munich?"

In unison, with the eager sincerity of a couple of five-year-olds, we replied, "Oh, we've never been."

It took us another hour before we figured out that we were in Munich, and that "München" was Munich in German—and *not* the present participle of an English word that means "to eat."

When we disembarked in Prague on November 3, 1998, in three-degree Fahrenheit weather, the coldest November on record (with, incidentally, the clothes we had packed for a move to a Mediterranean country), we put on every piece of clothing in our suitcases to avoid hypothermia. We were young. We were alive. We were in a country with socialized medicine. What could go wrong?

By 2000, Amy and I were living and traveling together: to Italy, France, Poland, Hungary, Ireland, Scotland, and England. As my second Prague spring rolled around, I was working as a producer for an English-language theater company and an editor at an English-language press. I rarely spoke to my family. I didn't yet have a cell phone, and with the time difference, it was often easier to leave each other messages.

After spending a weekend at a work retreat, I came home to find several messages from my mother. I picked up the phone and nonchalantly dialed the international number, listening to the now-familiar trill of the long-distance ring. When my mom answered, she asked to speak to Amy. It was an unusual request, but I don't remember reacting to it. I handed over the phone and began scrambling some eggs for dinner. When Amy hung up, I turned to look at her. Her face was gray.

"Someone's dead," I said quickly.

She nodded.

"It's Jeromy," I added.

She nodded again.

I looked back at my eggs, which had morphed into a strange yellow substance floating in a pan—for what purpose? I left them and went into my bedroom. Some amount of time later, Amy came in.

How did he die? I probably asked. Did he . . .

She didn't have to tell me any of it. Jeromy and I had spoken intermittently, and seen each other once when I had come home for a visit. On June 10, 2000, the twenty-nine-year-old poet checked himself into the Holiday Inn on High Street, east of Broad, in Columbus. He had a rope and enough heroin to kill himself. He used the clothing rod in the closet to tie his noose, then he shot up.

Jeromy did it two ways, to make sure.

I fell apart.

When my father died, I was falling in love. There was absolutely no room for the agony of death. With Jeromy's suicide, my sadness doubled down. Where life had been a rich adventure, it suddenly clouded thickly over and boxed me in.

FIFTEEN

I spent another few weeks in Prague trying to stay afloat, and then decided to move back to the States. My friend Jason wanted to start a fashion business with some money he had inherited after his mother's death. He asked me to move to San Francisco and work with him. Having absolutely no idea what else to do, I agreed to go.

Jason and I drove cross-country from Columbus. We took the southern route in the hope of avoiding any bad January weather. We drove down through Memphis and Nashville. We laughed at the faux Americana of the town of Texarkana on the Texas-Arkansas border. As we drove across Texas one morning, the sky was heavy with low winter clouds. They descended over the road in front of us until we were encased in fog. We drove on that empty stretch of highway, no more than ten feet of visibility in any direction. When we came out of the fog into an expanse of blue velvet sky, the red sand around us glittered in the sun. I looked in the rearview mirror and saw the wall of thick cloud behind me. It rose up as high as I could see, like the world was starkly divided in two.

The phone call came in to the house phone of my San Francisco apartment about a week before my cousin Danny's wedding. When I answered, it was my aunt Kathy, Danny's mother, my father's little sister.

"Norman's sick," she told me in a serious but hurried voice.

Norman, her other brother. My uncle.

"What?" I stuttered.

"Look," she continued, barreling through the bad news as quickly as she could, "he's going to look like Billy. That's what Ellen says." My aunt Ellen, Norman's wife.

"But Uncle Norman doesn't have it," I reasoned. "He doesn't have the murmur!"

"I'm just telling you what I know. They want us all to be prepared for what he's going to look like at the wedding."

Aunt Kathy's son, Danny, and his fiancée were getting married in June in Phoenix in an outdoor wedding. I had teased them that it was like planning an outdoor wedding in Minnesota in February. I had bought a new dress.

At the time when I received the call, Norman had already been sick for a year and hadn't told any of us. Only his wife and kids knew. He explained that it was because he simply hadn't been able to bring himself to tell his parents.

When my uncle arrived, surrounded by his family, he was gaunt and had my dad's illness's signature belly.

It all came out during a camping trip Norman took with his wife, his son Jeffrey, and his son's girlfriend's family at Lake San Antonio in central California. Norman had already started retaining lymphatic fluid in his abdomen, just like his older brother.

Norman wasn't prepared to get sick, even if he should have been. For one thing, he didn't think he'd have to prepare. He was supposed to be free of the gene that caused this very problem. After my dad died, Dr. Kricket sent him to a cardiologist at the University of Southern California for testing. They hadn't heard the telltale murmur that indicated the gene. So why would Norman think that the small bit of distension in his gut—if it was even visible—was anything more than a little weight gain, or just some bloating due to gas? And the little cough he had picked up? Well, he'd gotten an inhaler to help with that, even if it hadn't helped. In 2000, four years after Billy's death, Norman had gone on vacation with family friends who had pointed out that his legs were swollen. That hadn't convinced him either.

As Norman dove off a boat on that beautiful sunny day, he did not expect the shock of the cool of the water to suddenly restrict his airways. He did not think that when he landed, the fluid inside him would weigh him down. He certainly didn't anticipate that the fluid in his belly would restrict his diaphragm, the pressure of the water forcing it upward. The panic that ensued when he found he was unable to pull in a full breath likely exacerbated the pleural effusions—the fluid in his lungs sufferers of this illness all came to know so well—launching him into a full-blown panic attack.

Ellen and Jeffrey watched from on board the boat, calling out, "Are you okay, Norm?" "Dad, are you okay?"

Jeff's girlfriend's father dove in and guided Norman back up the ladder. Norman pulled away as he sat gulping for air, asking for a second to catch his breath. "Leave me alone!" he gasped. The group of near strangers stood around awkwardly in that small space, watching as Norman's breath eked its way back into his lungs.

Within a few minutes, he was okay, and the outing resumed. My uncle didn't get back into the water that day. Within a few weeks, Jeffrey and his girlfriend broke up—the product of graduation and being in your early twenties—and Norman let his son know the truth about the genetic lottery, which, like his brother, he had lost.

———

My aunt and uncle's marriage always seemed like a very romantic love story. From my perspective, my parents worked well together, but something about Aunt Ellen and Uncle Norman's relationship was different. They cared deeply for one another, but did not care as much about extending their compassion to others, or at least to Norman's family. That reverence for one another often left the rest of us distinctly on the outside. They loved their four children, but even my cousins (or perhaps especially my cousins) seemed a part of the everyone-else-ness of the world at large. This was most apparent in the way they teased us, never each other. It mostly looked and often felt like they told jokes to amuse each other at our expense.

To be fair, my observations were limited to family events. The few times I saw them away from our extended family, they were decidedly mellower. For example, in my twenties, when I stopped by their home for an unannounced overnight with a friend, I was surprised by their warmth and hospitality. Before we arrived, I warned my friend, "They are kind of mean, but they're really funny." Of course, they immediately invited us to join them for dinner. As we drove away the following morning my friend countered, "I have no idea why you think they're mean . . ."

If Uncle Norman was the funniest person in my family, Aunt Ellen was definitely the second funniest. The rest of us mostly appreciated their humor. Other times, for example when their sharp observations jabbed at our flaws and shortcomings, it was harder to appreciate. Ellen in particular used humor to deflect any real intimacy with our family. We, Billy's branch, were a family of the eagerly earnest, and were therefore easy game for Ellen's disdain. When her teasing was gentle, it was a pleasure. As it devolved into ridicule, however, you could only hope her focus had shifted to your sister, mother, or a cousin. She was like the family mean girl, so you pretty much always wanted her to like you.

My uncle was the classic middle child, full of resentment toward his family. Their sibling competitiveness was so fraught that when my father and I were putting together a family video for my grand-parents' fiftieth anniversary party, we had to count out the pictures one by one—Billy's family, Norman's family, Kathy's family—so that no one of the three children was disproportionately represented. I was assured that Norman *would* notice.

My aunt Kathy has pointed out that Norman's resentment toward his parents and siblings was a constant in the family dynamic. My grandfather took a promotion across the country when his children were teenagers. While Billy had graduated with the same group of kids in the same small New England town, my uncle was forced to transfer to an L.A. high school for his sophomore year. Then, when another job opportunity opened up for my grandfather, Norman was forced to transfer to yet another school, this one in Phoenix, for his senior year. His sister, Kathy, endured the moves as well, but each time the shake-up for her was less extreme, since each move coin-cided with beginning junior high and then high school. My uncle never fully forgave his parents for those moves.

My father and Aunt Kathy had always been much closer with each other than they were with my uncle. During family games, Kathy, the youngest of the three children, was paired with Billy, the oldest, while Norman, the middle child, was always paired with a parent. That dy-namic, and the jealousy and heated competition it gave rise to, carried over into adulthood: Kathy and Billy vs. Norman.

When Uncle Norman studied at the University of Arizona, he took a different direction from my dad, who had graduated premed with a chemistry degree. Norman, charming and funny, studied and then pursued a career in marketing. Two years after graduation, he was back living with his parents in Phoenix and trying to get a handle on his life. Both my dad and Aunt Kathy had met partners and married them. My dad had moved with my mom to Iowa to begin medical school. Aunt Kathy and her husband, a Vietnam vet and soon-to-be career army man, were preparing for a move to a base in Germany. Norman's life was the least settled, and his father judged him for it harshly. Sharing a home continued to corrode their already fragile relationship.

My uncle decided to take a trip to Tucson to visit old friends back at his alma mater. While he was there, they asked him if he wanted to meet a nice Jewish girl. Uncle Norman was game. That night he met Ellen. Six weeks later, they were engaged.

As Norman began to succeed in life, moving to California, working in sales and being promoted, first at one large aerospace company then another, there was always the sense that Billy and Kathy were somehow doing better. To their father, Kathy would always be the baby girl. My dad and his father spoke almost daily. My grandfather served as my dad's most trusted adviser and advocate.

My aunt and uncle decided to adopt when Ellen was told that she would require surgery to unblock her fallopian tubes if she ever hoped to carry children. She underwent the surgery, but soon after adopted Marcus. A year later, she gave birth to Aaron; two years later, she gave birth to Jeffrey. They knew they wanted a girl. They discussed going through in vitro fertilization and electing for a female, a very new technique in the mid-1980s, but then a different door opened when they learned about a two-year-old named Tiffany who had been given up for adoption. They called her Rachel, and with her, their family was complete.

In Uncle Norman's endless competition with my father, not having the faulty gene had been the one major way in which he had finally won. Now he didn't even have that.

For five years, my uncle believed he didn't have a murmur, so he didn't have the gene. His doctors at the University of Southern California had cleared him when he went in for a test with explicit instructions from Dr. Kricket. After he got sick, a ripple spread outward through the family. Without the murmur as the telltale sign of the disease, how could all the people who thought they were safe from this illness be sure that they were? All along, whether or not she had bad news for you, Dr. Kricket had offered a kind of security. She was a prophet with a very positive attitude.

"Yes," she'd say, "you have the gene, but you'll never get it like your father."

Or: "Yes," she'd admit, "you have the gene, but we will have answers by the time the next generation gets sick!"

Fortunately or unfortunately, the confusion was short-lived. Soon after Norman's diagnosis, Dr. Kricket asked my uncle to have his heart rechecked for a murmur, and this time it was clear that he had one. The USC team had simply missed it because it was so subtle. The dark upside to all of this was that my aunt and uncle had not spent five years waiting for Uncle Norman to start dying.

There was another upside too. When it came to his lifelong sibling rivalry, my uncle still had an edge—he just didn't know it yet. His victory wasn't going to come from not having the gene . . . it was going to come from not *passing on* the gene.

Although mapping a specific gene in the early days of the new millennium involved time-consuming and prohibitively expensive techniques, Dr. Kricket had made it a priority. She and her team had a theory based on the family members who had presented with a heart murmur, the anomaly they were continuing to use as a presumptive genetic marker. They believed our problem gene looked like it might be on one very specific chromosome: the X.

All humans have forty-six chromosomes. Each of those chromosomes contains, on average, twenty thousand to twenty-five thousand genes, therefore narrowing things down to one chromosome would be significant in helping to pinpoint which of those twenty-plus-thousand genes contained the variant that was causing my family members to die. When people and their reproductive partners pass the forty-six chromosomes of human DNA to their children, each partner contributes half. For every chromosome, each of us passes one chromosome out of the two possible chromosomes we inherited from our mothers and fathers. One of the chromosome pairs, among many other things, codes for gender. The chromosome passed by the father dictates the sex of the offspring. Every person in the world usually has two chromosomes that suggest whether they are physically female or male. Women have two X chromosomes, and men have one X chromosome and one Y chromosome.

Women, because they have two X chromosomes (one from each parent), will absolutely pass any child an X chromosome. Men, however, will pass on either a Y *or* an X. If a man passes on an X, he gets a baby that is coded for female. If he passes on a Y, he gets a baby who is coded for male. There are always exceptions—given how amazing and complex our genes are—but speaking generally, this is the process.

In our family, assuming Great-Grandma Mae was passing on a gene, the fact that it seemed clear that she passed it on to two girls (my grandmother Shirley and her sister, Norma) and one boy (my great-uncle Nathan) did not by itself prove anything. But the fact that my great-uncle Nathan had passed his gene to his daughter and *not* his son offered one small but valuable conclusion—namely, that the gene might be on the X chromosome. When Uncle Norman fell ill, Dr. Kricket had another male subject. Even more conveniently, Uncle Norman had two biological sons. My father had two biological daughters. Hilary and I both had murmurs. My male cousins did not. Again, this made the X chromosome a contender.

In the genetic lottery, my father, Uncle Norman, and Aunt Kathy were each poised to get one of their mother's X chromosomes. They had a fifty-fifty shot at receiving the X chromosome that my grandmother had received from her father, Rabbi Morris, or the X carrying a mutation, or variant, passed on by her mother, Mae. My dad and Uncle Norman both got the one that carried the variant, and Aunt Kathy got the other one.

It would not be easy to prove what Dr. Kricket was suggesting. Every time my cousin Jeff coughed, it was impossible not to wonder if the Seidmans were wrong. But if they were not wrong and the gene was on the X chromosome, Ellen's children would never get sick. For his line of descendants, the gene would stop with Norman.

———————————

My uncle didn't tell anyone he was sick because he had seen what my father's death had done to his parents, my grandfather in particular. Everyone had heard the stories about my grandmother standing over her husband, who had been weeping desperately nearly every night since he'd lost his son. My grandma had finally convinced him to go on antidepressants, which had taken the edge off his despair. But all of us saw a distinct change in our grandfather. He was quieter since his son's death and a lot less in the mood to teach us poker or play a game of Scrabble. My aunt Ellen and uncle Norman invented excuses for months about why they were going to skip Thanksgiving that year, why they weren't interested in a family Passover seder, always avoiding too detailed an explanation. The secret was easy to keep, given that they lived outside of Los Angeles, more than three hundred and seventy miles away from my grandparents in Phoenix.

When Uncle Norman contracted pneumonia at one point, it was the easiest and most truthful excuse they used. When my grandmother offered to come out and help while Norman convalesced, he and Ellen had to say no, encouraging her to stay in Phoenix instead.

My uncle wasn't just keeping secrets from his family—he was also keeping them from his doctors. He hoped to give them a chance to reach an unbiased conclusion about his prognosis. He thought it would help them catch something that, in four years, my dad's doctors never had. My aunt and uncle actively kept the fact of my father's death from Uncle Norman's medical team.

I wonder if this withholding of information, both from his medical team and from his parents, wasn't also a way for Norman to avoid the terrifying truth, as if, by never saying something, it wouldn't really be happening. In much the same way that forty years of silence had separated Mae and Nathan from Billy and Norman, this new yearlong silence might have been another way for Uncle Norman to turn away from the isolating and grim reality that awaited him.

SIXTEEN

As my grandparents stood before their son at my cousin Danny's wedding, their faces went gray. Three years after losing their eldest child, they were now poised to lose their second. My uncle Norman had the same gaunt look that my father had had. Three of his children and his wife, Ellen, flanked him like guards, or perhaps it was only Norman's vulnerability that made it seem like his family stood sentry around him that day.

I recall my grandparents dancing at the reception. My grandfather looked heavy in my grandmother's arms. She leaned into him as if her body weight was keeping him standing up. My grandmother's own mutated gene was set to claim yet another victim. It was her fault that her husband was going to lose another son. As I watched them, my grandmother in her bright blue pantsuit and diamond earrings, my grandfather with his silver hair, still thick despite his seventy-nine years, that dance was heavy and slow, more like a glorified attempt to remain upright.

From that point onward, my grandfather made a vow to his wife, his son, and himself: he would make amends for all the years that

Norman had felt underloved and underappreciated by his parents. He arranged for a trip with Ellen and Norman to Las Vegas. On that trip, he spent meaningful time with his son. He was gentle with his sick boy, helping him carefully maneuver around the slot machines, sitting quietly by his side as they played the games, the noise of Vegas never penetrating the peaceful bubble surrounding father and son as they healed their relationship, even as both their bodies weakened.

———

It wasn't enough that Shirley's second son's death was imminent. My great-aunt Norma, my grandmother's sixteen-months-younger sister, her lifelong companion, and most combative opponent, 2800 miles away in New Jersey, suddenly fell ill.

Norman was not named after Norma, according to my grandmother. Under Jewish law, you did not name children after the living. Still, I always associated Uncle Norman and Aunt Norma simply because of the similarities in their names. That January, Norma had begun to fill up with chylous fluid. Norma knew she had the gene. She had the telltale swelling and the murmur. Her doctor noted that the fluid Aunt Norma was accumulating was "very unusual." It was lymphatic exudate—that "pretty" light lemon-yellow-hued gravy called chyle.

A generation separated Norma and Norman, but their bodies were arriving at the same point at the same time. Norma was seventy-five and Norman was fifty-one. Similarly, a generation separated Mae and her son Nathan, while only a handful of years separated their deaths.

This reality served to bolster Dr. Kricket's assertion that the X chromosome was going to be a major player if she could track down the specific gene. The reason is a phenomenon called "X-inactivation."

If you Google "X-inactivation," you will almost certainly come across a bunch of pictures of adorable cats, which might have been exactly what our family needed to cheer us up back in 2001, if you could even Google adorable cat pictures back then. The reason for these search results is that a calico cat is a visual reminder that females have two X chromosomes.

Calico cats are so named because of their multicolored, checkered fur coats. For a brief moment, when an embryo that is chromosomally XX is in its zygotic state, both X chromosomes express. Imagine both X chromosomes flashing with light in every cell for just that short period of time. It's almost like the cells are having a little test run to see which X they like better. The chromosomes express themselves in this way only for a few moments, but in cats, that moment becomes forever etched in fur. If one of the cats' X chromosomes selects for black fur, and the other X selects for orange, a mixture of both in varying shades will remain. After that moment of dual expression, one X will package itself up and go silent while the other X carries on. This also explains why XY male cats are almost never calicos. In fact, all calico cats are XX females unless they have a chromosomal anomaly.

With all the other twenty-two pairs of chromosomes in our cells, both chromosomes in every pair express with very limited exceptions. Only the X chromosome likes to work alone. Some scientists believe it has to do with evolution. Both the X and Y chromosomes are big mutators. In almost every generation, some degree of mutation takes place in both.

Truth be told, they seem largely to be mutating to avoid killing each other off—you know, like in a marriage. One study of fruit flies allowed a male strain to mutate freely over twenty-nine generations. The female strain was bred repeatedly so that it never evolved and was virtually unchanged. When the twenty-nine-generations-into-the-future male fruit fly fertilized the "ancient" female, his sperm actually poisoned her. In human terms, that'd be like a man from today having sex with someone from the 1500s. Turns out all King Henry VIII needed was a man circa 2016 to take out his queens, and he could have avoided all the marital bloodbath.

So, perhaps as a tool of self-preservation, the X, unlike all the other chromosomes, has mastered the art of working alone. For a brief moment, her partnering Y must still code for things like testes and James Bond movies.

If the X chromosome's partner is another X instead of a Y, as in most females, she still won't want to share her efforts. After that momentary dual activation of both X chromosomes, only one X carries on while the other goes silent. Some scientists believe it is the "better" X that stays active in the cells, but there seems to be some degree of random mixture—a maternally inherited X remaining active in some cells and the paternally inherited X in others. Norma (like her mother, Mae) very likely had some mutated X's from her mother expressing in some of her cells and in other cells she had the X she inherited from her father, without the mutation. Or it's also possible that in the moment when both X's were active together, there remained that lingering residue of the silent X, like on the coat of a calico cat, that would play a role down the line in the life of an XX female. Either way, the probability is that Norma had a better chance of living longer, given that her body had, or momentarily had, another X from which to acquire information. In the cases of Norma and Mae, they each eked out an additional generation.

Infer whatever jokes you will, but the Y chromosome inherited by men is one of the smallest, least complex chromosomes a human can inherit. Once it has completed its task, it is packaged up and silenced, leaving the mighty X to carry on with her solo expression.

Nathan, Billy, and Norman, as men, inherited only one X chromosome. Because Nathan, Billy, and Norman's bodies had no healthy version of an X chromosome like their female counterparts, the illness in them seemed to come on harder and stronger. It was killing them faster and in more devastating ways. Dr. Kricket liked to remind Hilary and me of this by saying, "You don't have what your father had."

Meanwhile, our ever-spry and aging grandma Shirley would echo, "You'll live a long life like your grandmother."

Still, Norma's decline was rapid in the end. Her son-in-law Vinnie concocted rich smoothies for her to drink. It was the only food she could keep down. Her daughters sat with her every day in her hospital room. Like Uncle Norman and the others who preceded her in death, the biggest problem for Norma was an inability to absorb protein. When her medical team inserted a feeding tube through her nose and into her stomach, the hope was that her nutritional levels would improve, but the tube irritated her. She was in pain. Finally, she asked that the tube be removed, but by then the irritation was so great that it brought on an infection.

Norma died in April 2001 with her children, grandchildren, and husband, Charlie, at her side. The infection caused by her feeding tube was listed as the cause of death, but the chylous ascites was everywhere. Dr. Kricket brought Norma's body to Boston to conduct her autopsy.

––––––––––

With my uncle dying, my grandparents didn't want to go to Aunt Norma's funeral, so Hilary and I decided to go on their behalf. We flew to New Jersey the morning of the funeral. Although it occurred to me that I shared the gene that killed Norma, I was soothed by the fact that she had lived a full life. Hilary thought the funeral was a little more ominous.

I overheard a conversation that day, that at the time I just took as a little bit of gossip floating around among cousins: my cousin Valerie, Aunt Norma's oldest daughter, had had her spleen removed six weeks prior. I heard people discussing it across Norma's kitchen table after we returned from the cemetery.

Valerie looked great. It was clear that whatever had happened, she was back to her old self, fully made-up face, healthy, and now laughing about her recent medical scare. No one knew what had happened to the otherwise healthy forty-five-year-old. She had thrown up and the vomit looked like coffee grinds. When the doctors investigated her condition, they discovered her platelets had plummeted to near-fatal levels. Her spleen was grossly enlarged. Upon removal, she told us, it was "mushy." I'll admit, I didn't even perk up at the word "platelets."

By the time of the funeral, Val was healthy again, and her anecdote provided the background banter of another funeral. Her illness was just another medical mystery in a family who was absolutely steeped in them. At the time, no one connected Valerie's strange condition with the other strange condition that now had a growing body count. After all, mushy spleens hadn't been a problem for any of the stricken relatives.

So the next morning, without fanfare, Hilary and I flew back home.

SEVENTEEN

O ver the past few months I have been forced to look at you
from a new perspective," Norman wrote to Ellen on Valentine's Day, 2001. "I have to sit back and watch you plan for a not so distant future alone.

"I am now blinded by the sheer magnitude of the love I feel from you," he continues. "The warmth of that love quiets any fears or anxieties I may have for my own future."

The letter is a heartrending tribute, a testament to a twenty-eight-year love affair, coming to an end. "My life has a validation few people could ever hope to achieve," he concludes. "Validation based not on the love that I have given to you, but on the love that I have been so honored to receive from you."

He signed it, "Happy Valentine's Day. I love you, Norman."

My uncle grew sicker over the summer. He visited someone best described as a witch doctor. He was sent home with what looked like a Geiger counter and a bottle of vodka. By September, he still lacked any useful answers. On September 11, at two A.M., he woke up. He began frantically hyperventilating in bed.

"I can't breathe," he told my aunt, who had woken up in the com-
motion. She asked him if she should call 911. "No," he replied, "I
need you to hold me." So my aunt put her arms around him, holding my uncle and trying to soothe his labored breathing. Although
across the continent, planes were flying into the World Trade Center
towers, that is how my aunt remembers the events of 9/11.

Uncle Norman's doctors remained most worried about his protein levels, which had plummeted to near-starvation readings. He
was sent home with bagfuls of albumin, which had to be pumped
directly into his veins through a pick line in his subclavian artery,
located just beneath his collarbone. Aunt Ellen also prepared protein
shakes (happily made of actual protein, unlike the ones Norman's
uncle drank in the sixties).

In November of 2001, my grandfather was diagnosed with
stage-four colon cancer. Two months later, he died. I imagine he just
couldn't watch another of his children go before him. Uncle Norman went to his father's funeral wheelchair bound. My grandma
sat stone-faced beside her dying son, before the casket of her dead
husband, with her oldest child five years gone and her sister only
months in the ground. Aunt Kathy wept quietly at her side. My
grandparents had been married for fifty-seven years.

Uncle Norman's son Aaron was engaged, set to marry that July in
Los Angeles. Norman promised himself he wouldn't die before the
wedding.

My uncle's life, much like my father's, was built around doctors'
visits. My aunt Ellen tried to stay positive, but she didn't believe
he would make it much longer. When they married, they had both
signed living wills, agreeing that they would not be resuscitated if
doctors ever anticipated that they could not live without life support.
Hours later, Norman had torn his up, saying, "I don't want to be the

guy who pulled the plug and the next day they find the cure." But after seeing what his brother had gone through, he asked to go back to their lawyer to sign a new one.

The week before the wedding, my sister flew to L.A. to take my uncle to doctors' appointments and free my aunt to deal with coordinating the wedding. There was so much to do, and Ellen wasn't sure she could handle it all. Uncle Norman loved getting tapped—having a pump inserted to pull out hidden pockets of scar-tissue-locked fluid. Our father had loved it too. Tapping the fluid meant making room for airflow to thirsty lungs. It meant being able to breathe. Hilary took him every other day.

At the wedding, Norman slowly walked down the aisle with his wife at his side. It was the last long walk he ever took.

On October 14, 2002, Norman died. The cause of death on his chart read: "Respiratory failure due to massive chylous pleural effusions due to severe lymphedema."

EIGHTEEN

During my uncle's illness, I was living in San Francisco. My boyfriend at the time was an avid pot smoker. Once, I asked his best friend if it would be okay to tell my boyfriend that I liked him better sober. The best friend looked shocked and replied, "That would be like telling a fat person you'd like them better skinny."

As a once-heavy pot smoker myself, I knew I was being somewhat hypocritical, but I had all but stopped smoking. I had moved on to whiskey and denial. My father's death was firmly in the rearview mirror of my life, and I refused to let all the unknowns of a theoretical gene get in the way of my pleasant, if myopic, view of life. I was healthy. Other than my chronic low-platelet count that sometimes left me with mystery bruises, I felt healthy. Just like it's hard to make dinner when you're not hungry, preparing for an unhealthy future is often that thing on your to-do list just below dusting the tchotchke shelf and nose-hair plucking.

I focused with all my might on my healthy octogenarian grandmother. If she wasn't sick, I wouldn't get sick. I called and visited her regularly, mostly to feel her place her hands on my cheeks and say, "You'll live a long long life like your grandmother."

My grandmother's legs remained thick with edema. She wore heavy-duty support stockings that she pulled on slowly and with grunts. I never thought about all the effort she put into getting dressed, and although she was otherwise strong and spry, I chalked her difficulties up to old age. So if I was busy comparing myself to her, it shouldn't have come as a shock that one day I noticed my ankles looked swollen.

I don't suppose they swelled for no reason. It probably happened during a long walk or hike, which, for a mostly sedentary girl like me, weren't a common occurrence. One afternoon I had parked on one side of the Golden Gate Bridge and walked across it. I was alone, unemployed, and miserably entangled in an unfulfilling relationship. Amy, my best friend, who was still living in Prague, mentioned on the phone one day that when she pictured me in San Francisco, she always saw me walking across the Golden Gate Bridge. So one day I went. At one end of the inevitable two-way walk, I was so unimpressed, it was as if I suddenly understood the impulse people seemed to have to jump from this bridge. The pretty view aside, it was freezing and the car fumes mingled with the sour bay air. I hated the idea of having to recross to get back to my car, but didn't have a choice. My throbbing feet didn't either. Later that night, after I had collected the stoner from work, eaten someplace with a bar, and driven home, I peeled off my shoes, socks, and pants to see a visible "pouf" between calf and foot.

I moved my feet up the wall and watched a movie. The next day the swelling was gone. But over the ensuing weeks, it continued to make the occasional random appearance, changing in size from day to day, ankle to ankle.

At this point, I had seven years, three cities, several countries, and two oceans between my father's death and my current life. I knew that I had the gene that killed my dad. I knew that when the time came to have children, which was not going to be while I was dating the stoner, I should call Dr. Kricket. There was some talk that Dr. Kricket and others at her lab were working on solving this very problem. How, I wasn't sure. I knew that they would have to locate and map the gene, but all of that still seemed so abstract. I simply had the instruction "When I want to have a baby, call the Seidmans," which I stored in the back of my head along with "Smell milk before drinking" and "If bleeding, call 911."

A future marred by some vague, unknown genetic anomaly was tough for me to imagine in my mid-twenties. I continued to simply live my life, not really especially interested in the potential impact sharing this gene with my father would one day have on my life, but growing increasingly aware of the psychological wounds left by his protracted death.

When I think about my San Francisco years now, they are rife with long, lonely drives on California highways. I was twenty-five years old. Then I was twenty-six. I didn't have many friends except for the stoner, who only laughed at my jokes after I explained them. I hated him. I hated me. I was twenty-seven, then suddenly twenty-eight. I decided to visit a therapist, because that's what I'd been taught to do when the road ahead was foggy like a San Francisco afternoon. I had read in a magazine that you should shop deliberately for a therapist, meeting with many before deciding on one. It only made sense.

I met with a woman who reminded me of a newscaster with a perfect helmet of blond hair. She bobbed her head while I spoke. I gave her the abbreviated version of the shittiest things that had ever happened to me. When I was finished, her head had stopped bobbing and her mouth was agape. She suggested I start coming three times a week and go on a cocktail of medications. Preferring a traditional cocktail, like a whiskey sour, I tried out another person I found in the phone book, this one named Rebecca. She was six foot four and had an Adam's apple peeking out over her string of pearls. She was animated and seemed genuinely interested in what I had to say. I liked her. But as soon as I wrapped up my tale of woe, she looked at me with something almost gleeful in her eyes and exclaimed, "It sounds like you really have it all together!"

"Well, I wouldn't say *all* together," I cautioned.

"No, look at you. You've had such a time of it, and here you are, doing so well!"

"Well, I wouldn't say *so* well."

"Oh, sweetie, trust me; you really *really* have it together," Rebecca insisted.

But even I knew I really, *really* didn't have it together. I had a drinking problem. I had a stoner. I had no direction. I had a kitchen that smelled like cigarettes and three weeks' worth of dirty dishes. But maybe compared to some of Rebecca's other clients—hell, maybe even compared to Rebecca—I was doing really, *really* well.

Believing there had to be a middle ground between the lady who wanted to narcotize me and the lady who wanted me to become a mentor, I decided to keep on shopping. I listed all the shittiest things that had happened to me, one after the other:

Ages 18–22: *Father (slowly, horrifically) dies.*

Age 22: *Learn I have a genetic condition that may or may not slowly cut short my life.*

Age 22: *Fall in love with a heroin-addicted poet.*

Age 23: *Move to Prague to get away from heroin-addicted poet boyfriend.*

Age 24: *Ex-boyfriend commits suicide two ways (in order to be sure).*

Age 24: *Move back to America.*

Age 25: *Move to San Francisco.*

Ages 25–28: *Fall into an uncomfortable codependency with a stoner.*

Ages 25–28: *No friends.*

Ages 26–28: *No job.*

Age 28: *Starting to get worried about the swelling in my ankles.*

Age 28: *Having trouble with transitions.*

Having trouble with transitions.

The last one was the real reason I had sought out a therapist. As if the other items on my list weren't bad enough, I was having trouble with transitions. I couldn't go to bed at night. The clock would tick later and later. I didn't want to transition from awake to asleep. It felt too hard. But eventually, at some point, sometimes the next day, sometimes at four A.M., sometimes at five P.M., I would go to sleep. And then I wouldn't wake up. I didn't want to. It wasn't about feeling tired. I mean, I *could* wake up. I just didn't feel like it. So I'd go back to sleep, waking up briefly only to decide it was easier to roll over again. Eventually, probably because I needed a cigarette, I'd finally get out of bed. But other than maintaining that tired cycle of addiction, nothing motivated me. I was caught in whatever state I was in. If I was eating, it was really hard to stop. If I wasn't eating, it was really hard to start. If I was sitting, why stand? If I was standing, why sit? Playing or not playing computer solitaire. Drinking or not drinking whatever alcoholic beverage was closest. I closed bars because I couldn't leave. I became practically agoraphobic because I couldn't go out. I couldn't get a job because I didn't have a job.

And then there was this epiphany: *I didn't have any wishes anymore.*

Like everyone, I had always loved to talk about what I would wish if I ever found a genie or a leprechaun or a magic wand. After wishing for a hundred or a million wishes, I'd rattle off my list to a parent, my sister, a dog, whoever would listen, and it was never a bad list either: amazing chocolate; friendly mythological friends; a scratch-n-sniff sticker that was also fuzzy and glittery; a neon-pink Forenza sweater from the Limited, circa 1985; a date followed by a long happy marriage to Ethan Hawke as the kid in *The Explorers,* Ethan Hawke as the guy in *Dead Poets Society,* or Ethan Hawke as Troy from *Reality Bites* (decade-dependent); one of those huge blue bows Mary Ingalls pulled off on *Little House;* an endless *Little House* marathon; the lipstick color Samantha Mathis wears in *Pump Up*

the Volume; "Careless Whisper" by Wham! to come on just as I have
gotten up the nerve to ask Jeff Heiny at Jill Barnett's bat mitzvah to
dance with me; and well, I could always wish for world peace, riches
for the poor, wisdom, health, and a long and happy life.

But as I was about to start therapy, I realized I had no wishes at
all. There had been something about the catastrophic loss of my father, followed by Jeromy's suicide, not to mention my own uncertain
future, that had just depleted the world of edges and color. I was fine,
always. Neither sad nor happy. Just fine. With a low-grade boredom ever present like a tan cardboard ribbon through the flavorless
marble cake of my existence.

"Death and taxes . . ." I used to shake my head and lament like a
seventy-five-year-old vacuum-cleaner salesman. "That's all there is."
Why make a bed you're only going to mess up later? Why bother to
live if at some point you die?

The boredom that engulfed me during those years was suffocating. It felt like a warm wet blanket covering your nose on a humid
day. It smelled like the velvet ropes in line at the bank when you are
four years old and standing there with your mother. It looked beige.
My boredom overwhelmed me. When I considered getting a job,
contemplated going to a party or a movie, traveling to someplace
exotic, or even getting the most amazing sticker ever designed *including a glittery scratch-n-sniff flower,* it all just sounded . . . meh.

When I finally found the right therapist, she offered up an observation that I took to heart: boredom, she believed, was a form
of anger. It was anger on hold. Anger on autopilot. According to
my chosen and carefully shopped-for therapist, tall with a perfectly
bouncy head of dark brown hair and dressed in comfortable casual
wear, boredom and anger went together like peas and carrots, she
said. I thought that sounded absolutely right.

NINETEEN

With my mental health under the care of professionals, I agreed to a trip to visit Dr. Kricket. More specifically, Dr. Kricket had responded to an e-mail I had offhandedly written to her one day, asking her if I should be worried about my ankles. She came back with an offer of a free plane ticket to Boston. She would incorporate the visit into her study and could use grant money to bring me in. Of course, I was lethargic, which made the idea of travel challenging. But not going to Boston seemed as *whatever* as going to Boston. So, free-ticketed, I opted to go.

I hadn't been to the lab in five years, since just after my father died and my sister and I had gone with our grandmother after my faux-college-graduation ceremony. I had never actually visited the Seidman lab. I had always gone to an echo lab at Brigham and Women's Hospital, just off the Green Line, near the art museum.

Barbara, the wonderful strawberry-blond nurse, scooped me up in the hospital lobby and hugged me warmly. She delivered me to a room where I dressed myself in a gown and then had an ultrasound. Faranak, the same echo tech who'd conducted this very procedure on me three times now, carried on the friendly chatter. Once again, she had slipped me into the day's crowded schedule as a favor to the lab, but more likely as a favor to Barbara or Dr. Kricket, both of whom doted on her affectionately before and after the test. "Faranak's our girl," Barbara said, not for the first time, giving her a warm squeeze. "We love her!"

"Me too," I agreed. It was true. I loved them all. They oozed kindness and it wasn't the fake kind I had grown used to in my life in San Francisco, where intimacy came strangled in hard liquor.

After the test, I waited for Dr. Kricket to make her way in and give me an update. This time, when she walked in the room, she wasn't alone. A pretty white-coated doctor who looked about my age accompanied her.

As the lead singer in a short-lived college band that played in public only six times, I was once greeted by a fan with a hollow moony glaze across her face. It felt a little crazy, to be honest, being stared at like a celebrity. Having more often been the fan in similar situations—as the girl who really does get a little choked up when she runs into local newscasters—I totally understood the sentiment.

I mention this because the first and only time I ever met Meredith Moore in person, this is exactly what it felt like. When Meredith, a tall, dirty blonde with dusty freckles and glasses, approached me in the examination room, she was breathless and giggly; at one point it seemed almost like she wanted to touch my hair. Frankly, I wouldn't have been surprised if she'd asked for an autograph.

Truth be told, there was something just a little bit glassy about everyone's eyes that day. But Meredith Moore had it bad. It was like she knew me. She held my gaze and laughed a little too appreciatively before I finished my sentences.

There was a simple explanation for all of this. She actually did know me—in some ways better than my mother, or even better than *I* knew me. She knew my genes and she knew them intimately. Meredith was the postdoc fellow who had been studying my family's genes, and she had been doing so for more than a year. She had been the reader of my DNA, human RNA. She had been reading my genes to see how I'd been made. Now here I was, the living embodiment of that recipe book she had been staring at under a microscope.

Seeing me, even in my full-scale late-twenties depression, was a very real moment for her after all those months of study. What was more, she had news for me. It was big news. In fact, *I* should have been fangirling all over *her*. Meredith had had a breakthrough. She was about to tell me all about it—one of fourteen people in the history of the world who could truly appreciate this breakthrough.

And even in the haze of my depression, I understood the significance of it: Meredith Moore had mapped our gene.

Meredith Moore came to work at the Seidman lab for her postdoc in 2001, in an apprenticeship of sorts. After she left that job five years and one mapped gene later, she ran her own genetics lab for three years. Today she is a full-time research administrator and manager of the animal welfare program at the University of Texas. She has a cat she inherited from an old friend who passed away, and she recently gave birth to her first child.

Meredith and I are the same age. Something about this felt immediately important to me. How did the lives of two such different twenty-nine-year-old women intersect in that small research lab in 2004: one who'd just finished searching for a therapist, the other who'd just finished searching for a family gene?

Meredith grew up in Pennsylvania, studious and self-admittedly borderline OCD. After high school, she got enough scholarship money to have her pick of out-of-state schools. She chose the University of Delaware, which had the chemical engineering program she liked best.

The first half of her freshman year constituted about as rebellious a phase in her life as she would ever have. Between parties and socializing, her GPA started to fall. To keep her scholarships, Meredith had to eke out 60/1000ths of a point on her GPA on her last report card of the year. Her mother celebrated her daughter's accomplishment with a cake that read simply *60/1000ths.*

In college, Meredith's drive to save the world led her to major in chemical engineering. She soon realized that the kind of organization and planning skills required to do research appealed to her OCD side. She became a protégée of a professor who was studying the function of mitochondria in a cardiac cell. Though she had always thought she would prefer to work on problems she could see with her own eyes—like a beating heart—she found herself spending a lot of time looking through a microscope. Much to her surprise, she found she liked working with the DNA in a heart cell. Looking under a microscope, she was amazed to find that the cardiac cell kept beating outside of the body. But what hooked Meredith, totally blew her mind, was that it wasn't just the cardiac cells that kept beating . . . individual parts of the cell—like the mitochondria—also beat.

Still, Meredith wanted to make sure she was helping actual people. Prompted by her mentor, she shared her research with the Seidman lab and was immediately hired. From day one, her job was to create a version of something called a "DNA microarray." A microarray is basically a spot of DNA attached to a solid surface, typically a glass slide, that can be analyzed by a computer. Back in the early aughts, scientists still had to slog through the tedious process of actually sticking DNA sequences onto slides before they could get around to studying them. Meredith arrived at the Seidman lab and was tasked with figuring out the best way to attach DNA to glass. (I, on the other hand, was at this time somewhere in California listening to Euro rock on repeat and wishing for a new season of *Sex and the City* to come out on DVD.)

But changes in genetic medicine were already coming about

quickly, and around Meredith's one-year mark with the Seidmans in 2003, companies had started manufacturing microarrays that labs could buy. In other words, you tell them what DNA sequence you want, and like a version of Blockbuster for science geeks, that sequence will simply show up in the mail. No muss, no fuss, and *way* less expensive than paying a whole year's salary to a postdoc fellow playing hit or miss with slides.

So, Meredith told me, "That's when Dr. Kricket said, 'I have this other project . . .'"

It was us. We were her other project.

Little did we know, once Meredith was on our case, everything would change. She applied her outstanding attention to detail and endless gumption to dig up files for Patient A, my great-grandmother Mae; Patient B, my great-uncle Nathan; Patient C, my father Billy; Patient D, my great-aunt Norma; and Patient E, my uncle Norman. She similarly labeled the other nine of us who were still living and relatively healthy and gathered our medical records. Meredith compiled photographs of us and constructed an intimate genetic story of a family. She even found a photo of Mae's mother, Ester, with the telltale swollen ankles. Eventually, Meredith began to feel close to us, especially to my cousin Phyllis, who lived in a suburb of Boston and came in to meet with her on a fairly regular basis.

Our case motivated Meredith. We needed help and she wanted to give it to us. Science was still a long way from "curing" genetic diseases, but we lacked nearly every answer. Although much had been learned since my father's death, very little of it was practical information.

Up until 2003, treatments for genetic diseases were still largely superficial. While no genetic diseases could be cured outright, some could be held at bay with medication. In rare instances, treatments could trick the body into something resembling a cure. But in order to even begin to look into those possibilities, we had to better understand the mechanism that was causing our illness. We knew a thousand things that *weren't* causing it. We had no idea what was. The first step toward getting answers was finding the gene causing the problems. Meredith made mapping our gene her top priority. Her drive to connect to patients, to actually *feel* like she was helping people, was finally being nourished. And my family would reap the remarkable benefits.

TWENTY

You can't actually see DNA. Were you wondering? Were you asking yourself, how would Meredith Moore actually look at my DNA? Did she put a cell under a microscope and look at a floating little ladder-shaped double helix, maybe with a small cartoon gorilla smiling and waving from its rungs? The answer is, obviously, no. When scientists look at DNA, they look at colors. Researchers have to copy a single piece of DNA, made up of a nucleotide sequence, and just keep copying it over and over. To copy DNA, scientists heat up the two strands of DNA until they can be pulled apart. They use an enzyme to copy the information on each of those strands before putting them back together again. At that point, they have two double strands of the same DNA. They heat up those strands and separate them. Use that good old enzyme to copy them, and put them back together again. Then they have four strands of the same DNA. These strands are copied until finally there is enough so that when a bit of dye is added to the gobs of DNA copies, those copies will turn a certain color when exposed to different kinds of light. What these researchers are actually "seeing" when they look into the microscope is the dye that they have stuck onto the nucleotides. That light makes the dye glow in one of four colors: red, green, yellow, or blue.

DNA is made up of only *four* nucleotides—one for each color that you see. That's wild, right? There are only four of them, just showing up in some different order so that they are making the recipe for "eyeball" or "sit bone" or "hamster paw."

But it gets even more nuanced. How can it get more nuanced than *four*? you ask. Well, each of those four nucleotides has a buddy, or a significant other. The four nucleotides—cytosine, guanine, thymine, and adenine—are paired cytosine and guanine, and thymine and adenine. These are "base pairs."

By looking at the color they produce under a specific light, Meredith combed through my family's DNA and "looked" for our gene, one base pair at a time. Let me be clear here. There are approximately 3 billion base pairs in the twenty-three chromosomes of the human genome. Around 153 million base pairs are in the X chromosome alone. And there can be as many as 27,000 base pairs on a single gene.

Meredith was getting ready to look for our mutation somewhere on one of ninety-six genes. A mutation is basically a different or unexpected base pair (from anyone else's base pairs on that same gene) showing up *anywhere* along twenty-seven thousand repeating base pairs. If I had been in Meredith's shoes between the ages of twenty-seven and twenty-nine, this plan of attack—to simply read meticulously through thousands upon thousands of rows of the same two annoying couple of base pairs—would very likely have left me hiding under the nearest rock . . . or bottle of Jameson's. But Meredith was ready for the challenge. She was excited by it.

She immediately started looking for differences in the X chromosome of one male carrier and one female carrier against the X chromosome of a genetically unrelated member of the family, in our case a spouse of someone affected. She might have used, for example, me, my grandmother, and my uncle Norman's wife, Ellen, who is genetically unrelated to either of us.

That's how Meredith got started: scanning colored traces looking for differences in the X chromosomes, specifically differences in colors, since those were what she was looking at. If she saw something promising, like a difference between a nucleotide sequence on a gene in Ellen's X and a nucleotide sequence on a gene in my and my grandma's X chromosomes, for instance, she would then look at the same gene of every single carrier. If we all had it, she then had to make sure it was also *absent* from every single spouse and additional *unaffected* family member.

Meredith found thirty possible gene candidates out of the ninety-six genes she was studying. Those thirty candidates each had a code that was subtly different from the normal sequences of unaffected family members. Perhaps these different nucleotides directed the synthesis of protein somewhat differently than the "normal" ones.

So that's when her real work began. (Right? I know.)

Meredith's methodology at this point was pretty simple: start with the shortest genes. A gene is no bigger than the number of queued-up nucleotides that compose it. She picked the ones with the fewest nucleotides in order to knock those out first. Almost immediately, she stumbled upon a variant that segregated perfectly in that first set of samples. In other words, that variant showed up in one of mine and my grandma's genes and not at all in Ellen's. Then she checked a few more carriers, perhaps my cousin Valerie, Aunt Norma's daughter, who also had the murmur, and my sister, Hilary. As she continued to compare it to people with both affected and unaffected X's, she continued to get the result she hoped for.

What was especially exciting about this discovery was that this gene was unknown. In other words, it had no known function. If Meredith, and by extension the Seidman lab, could prove that this gene was linked to our illness, leading to a catastrophic breakdown in the lymphatic system, it could prove to be an enormously valuable breakthrough for science, another puzzle piece latched on to the mysterious human genome. But the first crushing blow to this hypothesis came when one individual—my great-uncle Yussy, who was by then well past the age when the disease should have struck him down, turned out to have that variant as well. It turned out that that variant had just randomly worked out.

Meredith used this experience to guide her in her testing. From that point forward, when she found an anomaly between my grandma and me, and my aunt Ellen, she'd start by looking at the genes of the "normal people."

Remarkably, Meredith honed in on the correct variant, or specifically four variants, all clustered in the same region of one gene, relatively early in her testing. The variants were on one of the short ones. In order to be thorough, she had to continue to lumber through each of the other twenty-nine genes, proving over and over that carriers had it and noncarriers didn't. Like that early variant that had at first seemed so perfect, odds were she might have come across a second, or even a third potential variant that segregated. In that case, she would have had to come up with another way to narrow down the focus even more.

But as luck would have it, no other mutations appeared, and the Seidman lab flew me to Boston to give me the good news. I would love to say I was immediately launched out of my black-hole life. I'd love to think that the news wowed me and shook me to my core.

Because as Meredith Moore gave me her biggest fangirl smile and told me about her momentous accomplishment, she understood that she had done something meaningful and good for my family that would positively impact all of our lives. Even though she didn't discover a new gene—she added to already existing information about a known gene, which was not something to sneeze at—she had done something monumental for our family. She had done what no one had been able to do in the five years my father lay dying: she had given us an answer that with any luck would lead to more answers.

It occurs to me now that if I had been myself in those days, if I had been halfway interested in my own life, I might have swept her into a bear hug and wept for joy. Instead I got back on a plane and returned to San Francisco. My ankles still swollen, I likely turned on computer solitaire and continued to have no idea what to do next.

TWENTY-ONE

In June of 2005, for lack of a better idea, I moved to Brooklyn. On a recent visit to New York City for my twenty-ninth birthday, a good friend had mentioned that come fall, she would be looking for a new place to live and that I could move in with her. Another friend had shown me the best websites for job hunting. I needed something to break me out of my funk and into some semblance of a life. Remarkably, this worked. When I told my therapist my plan to move, she had been wary. She suggested I wait six months to make sure it was what I really wanted to do. I am only partially kidding when I say that I think she might have enjoyed my twice-a-week, out-of-pocket checks. But I waited out those months, and then I drove back across the country, leaving my car in Ohio and taking a rental the rest of the way to Brooklyn. My lack of health insurance, and the increasing bloat in my ankles, made living four hours by car from Boston and Dr. Kricket very appealing.

Immediately I loved New York. I loved the energy and the palpable sense of possibility. Even more, I loved my friends, with whom I immediately fell into step. I got a job working for a small independent film company, and my friend Lisa and I got an apartment together.

Dr. Kricket arranged for me to meet with a blood doctor at New York Presbyterian regarding my exceedingly low platelet count, or "thrombocytopenia." My platelet count had been hovering somewhere around 60,000 since college. Normal is around 150,000 to 350,000 platelets. This hematologist, a colleague of Dr. Kricket's from her medical school days, was going to use a brand-new machine to test my platelets. I was told that Matt Lauer had been the first person to ever get tested with this particular machine, on live TV. At the time, the machine was the most high-tech imaging machine available. It could investigate blood flow, including the manufacture and function of platelets, better than any other machine.

The test was part of a study related to a long-shot theory a colleague of Dr. Kricket's had about the function of platelets. Dr. Kricket asked that I be included in the study, almost exclusively as a favor, since very few of my symptoms fit the criteria. The research team was looking for something called "supersticky platelets." According to their brand-new theory, some people struggling with chronically low platelets simply didn't need normal platelet levels because the ones they had worked so well. They were supersticky, the clotting champions of the world. Hilary and I had platelet levels that made us contenders, however unlikely.

I changed into a hospital gown and climbed into the strange spherical machine that had once held Matt Lauer, but I didn't understand the study or my place in it. All I knew was that my ankles were swollen and Dr. Kricket said to go, so I went.

With goodwill as my only health insurance since I'd graduated from college, I brought the team conducting the study some cookies. I met with the doctor, who glanced obligatorily at my swollen ankles and muttered a disinterested "uh-huh." I leaned forward in a seated position, holding on to two bike-handle-like contraptions. Two X-ray strips encircled my body. One strip ran around my body lengthwise and the other crosswise, but unlike one would be in a traditional MRI machine, I was not wholly encased. A tech came in and pricked the crook of my arm to prepare an IV port. The mood was light. I was certain that we were going to find the answers to all of my questions, despite the fact that no doctor in the fifty-plus-years of treating my family illness could tell us what was going on. What can I say? I'm an optimist.

The hope machine noisily shifted my body forward and backward and side to side, like a slo-mo version of the flight simulator Lea Thompson battles in the 1986 movie *Space Camp*. Dye was added to my blood through an IV. It tasted like I was drinking and subsequently peeing metal, a surprisingly common hallucination for anyone who has undergone an MRI with contrast. The dye helped the researchers visualize my blood flow. Something else was injected so the researchers could watch how my bone marrow (where platelets are produced) functioned, as well as how the platelets themselves behaved.

After thirty minutes, I was released, then sent to change back into my street clothes and be on my way. No one called for several days, so I called them. No answer. Two weeks later, an intern called me back.

"Everything looked good!" he said cheerily.

"What do you mean?" I asked.

"It looks good. Your blood looks good."

I thanked him, although I wasn't sure why, and hung up. I called Dr. Kricket for further clarification.

"Well, your platelets aren't supersticky," she admitted.

"So what do you think is wrong?" I asked.

"What do you mean? You're anemic. A lot of people are anemic." She'd resorted to her old answer. I'm not sure how knowing about the function of my platelets would have impacted my future one way or another, but it sure would have made a difference to have finally had a positive test result, good or bad.

Dr. Kricket didn't say it, but I knew that there were so many unknowns about my family's illness that grasping at straws was almost as good as following actual leads. Genetic illnesses begin in places that we cannot see, with damage often originating at microscopic levels. Our genome has at least three billion places to look at and interpret. It's like digging through a haystack for a needle when you don't even know if you're actually looking for a needle. Every strange object researchers find, they have to consider. It's a slow, tedious process, and when all of this was happening, that process was at its slowest and most tedious, because all of it was new.

―――――――――

Several weeks later, I went to another hematologist for a second series of tests, including a complete blood work-up, to see if the mystery of my platelets could be solved. When the doctor came into the room, he shook my hand, and then he looked down my pants. As a blood doctor, his jurisdiction definitely did not extend to my underwear. Yet he pulled up my hospital gown, lowered my pants zipper with a quiet "May I?" then pulled back the elastic band of my colorful Hanes all-cotton hipster briefs.

I understood what he was looking for because when I met him, I'd explained that I had a rare mutation on the X chromosome that led to an unnamed condition. The X chromosome is a famous chromosome. First of all, of all the twenty-three chromosomes, it is only one of two with two names. The X chromosome is also named "chromosome 23," but that's its *boring* name, and the X chromosome is a lot of things, but boring is not one of them.

It's like the Marilyn Monroe of chromosomes—Norma Jean, to those who didn't know what she could become with a little red lipstick and a lot of peroxide. I might be a little biased toward the X because that's where my mutation lives. Of course, if you have Down syndrome—which involves the occurrence of a third "chromosome 21"—*that* might be your most famous chromosome . . . your Tom Cruise of the chromosomal world, if you will. Or if you have Huntington's disease, you might feel like its location on chromosome 4 makes it Mick Jagger to the other twenty-three chromosomes' Ronnie Wood. But the truth is, the vast majority of us wouldn't notice any single chromosome playing a day-to-day role in our lives. But chances are, if you know a chromosome, you know the X, because we all have one, and it does a lot.

When the hematologist looked down my pants, he was probably looking for a vaginal anomaly. I don't have one. The hematologist was misguided in looking for one, but possibly, as a blood doctor, he slept through the genetics portion of his medical school studies, which I'm told is a really short one anyway.

The general consensus is that one single chromosome can code for multiple outcomes. In other words, no chromosome does only one thing. And X-chromosome–Marilyn Monroe, because she is *so* popular, is actually incredibly prolific, genetically speaking. What's more, science knows a relatively good amount about the X chromosome, which is arguably very little, but in the world of chromosomes is a huge amount. We know that a gene on the X chromosome codes for seeing colors. Another gene codes for the size of your cornea. Another still has a direct correlation to myelin, the coating that protects the neurons that make up your nervous system. There are around two thousand genes on the X chromosome. Each gene, in large part, is coding for its own thing. Occasionally, they work in groups. However, just because a guy is color-blind because of a mutation on his X chromosome doesn't mean there is anything wrong with his private parts. They just aren't related, at least not like that.

Meredith Moore had mapped my family gene on our X chromosome, and she saw something interesting. Our mutation was on the very same gene that sometimes has a mutation that serves as a marker for asthma. Since our symptoms were nothing like those of asthma, at least on the surface (lymphedema and starvation versus shortness of breath) it would seem obvious that this gene had other jobs outside of assisting with breathing. At that point, however, we weren't certain what those jobs were. What we did know was that if our mutation had simply chosen a different *part* of the gene to mutate, we might have all merely had a good old-fashioned case of asthma. But it hadn't chosen a different part. Not by a long shot.

I was visiting the hematologist on the off chance that he could tell me the cause of my low platelet count, which, along with my swollen ankles, remained an unexplained anomaly. During the autopsy of my great-aunt Norma, Dr. Kricket did not find low platelet counts or other physiological factors that might contribute to low platelet counts. And the hematologist, in addition to finding nothing remarkable in my pants (I can say that comfortably because I'm a woman), found nothing remarkable about my blood. I left that day with another diagnosis of "I don't know" and a phone call from Dr. Kricket reminding me that I would live to be an old woman just like my grandmother.

Except my grandmother, like *her* sister, didn't have a low platelet count. *My* sister and I did. So maybe our platelet count was something else entirely. As siblings, Hilary and I shared a lot of genes. Since no one else had the platelet issue that we knew about—I had by this point forgotten about Valerie's post-funeral mention of hers—it was possible that it was unrelated to the gene that had killed so many people in my family. Hilary and I did have a propensity for bruising, but we both had normal periods. When we cut ourselves, the bleeding would stop in short order. Platelets are the part of the blood that clots, or turns it from a liquid to a solid when it meets oxygen so that you won't bleed out from a hangnail. It was clear our blood could and always did clot normally.

Regardless, our platelets were going to have to wait. Soon after the results came back inconclusive from the hematologist, my phone rang. It was Hilary, my thirty-one-year-old sister, calling me from a bathroom stall, 530 miles away at her office in Columbus.

"Holy shit," she said, the echo of those miles sounding in every one of her words. "I'm pregnant."

TWENTY-TWO

"You're what?" I asked. It was my only available response.

"I'm . . . The double line on the pregnancy test . . ." she stammered.

"You can't!"

"I know!"

Hilary had been in a serious relationship with a man she was hoping to marry. Ever since she'd entered her thirties, babies were on her mind. Although they used birth control, unluckily this time it had failed them.

Hilary's regret was nearly palpable, especially as the boyfriend made it clear he did not want her to have this baby.

She reached out to the Seidman lab. Dr. Kricket told her she could have the embryo tested for the gene. It wasn't ideal, but by twelve to fifteen weeks, doctors would be able to tell whether or not she had passed on the gene. They could also find out the gender of the baby. If the baby was female, from what we understood, there was great potential for a long and healthy life. Aunt Norma had been in her mid-seventies when she had grown catastrophically ill. No one mentioned my great-grandmother who had fallen ill at thirty-five and died at the age of fifty-five, and her long-term stays at various hospitals, unable to breathe. Sometimes, it's easier to focus on the positive.

Over the course of several tear-filled days, Hilary wavered.

I remember at one point saying emphatically, "Hilary, imagine standing over the bed of your dying son."

If our grandmother had known the sentence she was imposing on her children, would she have had them? What if she could have made that early connection between her mother and brother? What if she could have seen in her own swollen ankles a future in which she'd be hovering over the swollen bodies of her two boys? Although stoic, my grandmother stopped crying after Uncle Norman's death, as she barricaded herself behind indestructible emotional ramparts. She knew better than to blame herself outright for their deaths, but that her gene had killed them was always on her mind.

Having seen the brutal death of my father, I felt it was my duty to do everything I could to keep this gene from becoming a scourge, overtaking who knew how many generations. At one time, diseases like cystic fibrosis had only impacted fourteen people, like ours did. That number had grown to the hundreds of thousands over the course of thousands of generations, with even more mutant variants expanding from that first strain. You can't stop a gene from spreading when its impact is that large. But we could! Hilary and I and two of our cousins who might still have children had a chance to decide to *not pass our gene*, even if the details of such a plan remained vague. Dr. Kricket had told us to let her know when we were ready to bear children and everything would fall into place. Of course it wouldn't be easy, but the terrifying alternative suggested untold generations of lymphatic swelling and starvation.

Hilary, pressured by her boyfriend and terrified she'd one day have to tell her child that she'd made a choice to carry to term a potential ticking time bomb, terminated her pregnancy in its earliest days. Soon after, her boyfriend left her.

Hilary does not go to movies after they've started. She claims it's because, when she was a kid in a darkened theater, she once accidentally sat on a large, bearded man's lap. I don't know that she needs such a precise anecdote to support her decision. The truth is, Hil has never liked to stand out in a crowd. I always thought how unfortunate it was for her that, given this trait, she should be a pretty six-foot-tall blonde.

Despite her attempts at invisibility, though, Hilary has always been hard to ignore. She was the star player on her tennis team, our high school's yearbook editor, and a news anchor on her university's local TV channel by her senior year. When we were kids, she organized and supervised all of our games, like a good big sister. Maybe it's because she's the elder child, but even now Hilary is always somehow quietly, and as far as I'm concerned, thankfully, in charge.

Having children was always in Hilary's plan. When her boyfriend broke up with her, she was devastated. Over the next three years, she dated like it was her job. When she met Brian, a landscape architect/urban planner who lived in Cincinnati, an hour and a half south of Columbus, she embraced the commute. Ten months later, they were engaged.

After Meredith Moore finished mapping our gene, Dr. Kricket had formulated a plan for safe reproduction. By 2007, our father had been dead for ten years. Hilary and Brian decided to consult Dr. Kricket. At the time, scientists could already read DNA well enough to locate and select for or against certain mapped wayward genes.

Hilary was told she could undergo in vitro fertilization, or IVF. IVF is the process by which a human egg is manually injected with

a human sperm in order to create a viable embryo. The term "in vitro" itself means "a process performed outside a living organism." In
1978, Louise Joy Brown became the first "test-tube baby" to be born,
having developed as a fertilized egg for several days before doctors
surgically implanted her in her mother's uterus, about five days after
fertilization. The five-day mark is typically when a viable egg reaches
something called "the blastocyst stage," meaning it has an inner-cell
mass able to develop into an embryo.

What's remarkable is that once a fertilized egg has reached the
blastocyst stage, its DNA is already intelligible. It is *readable*. Scientists need only one cell from this tiny collection of five-day-old cells
in order to have an incredibly detailed genetic recipe for the DNA of
the unborn person these cells might eventually make up.

It doesn't get much more sci-fi than that. Before a person is even
a person, genetics already makes it possible to anticipate more facts
about that person than even his or her own parents might know by
a first birthday! Certainly when we're considering the health and
wellness of an unborn baby, this fact has the potential for stunning
impact.

When Hilary and Brian married on Cinco de Mayo, Brian already understood the role IVF would inevitably play in the lives of
his future children. Luckily, he was a full-blown science geek, tech-
savvy and wholly unafraid of what Hilary had suggested to him: that
whether or not she became sick with the gene, their children could
be spared. Although having children was a priority for him, he felt
that Hilary had formulated a sound and viable strategy for safely
having those children.

One month later, Hilary and Brian met with a fertility specialist and explained her circumstances. The specialist recommended a
lab in Detroit run by Dr. Mark Hughes where her fertilized eggs
would undergo testing, five days after fertilization. The embryos at
the blastocyst phase that were not compromised by the gene that

had killed our father would be the ones that might potentially become her children.

Dr. Hughes is one of the most well-known and well-respected genetic embryologists in the world. He has been fighting for the rights of parents to give birth to children who are genetically healthy for a long time. The fact that he should have to fight for these rights truly seems to baffle him, as he's made clear every time I've seen him interviewed on television.

In 1997, a scandal forced Dr. Hughes to leave the National Institutes of Health, the same government agency where my great-uncle Nathan had been a patient for ten months in the early 1960s. The strong regulations that control most gene-based research, some might argue, provide ethical protection. When Dr. Hughes accidentally selected and implanted an embryo with cystic fibrosis in the wife of a couple who had sought his services in the hope of accomplishing the opposite, word got out that he had been using government funds to perform procedures of this kind for years. Cystic fibrosis, as I've mentioned, is an incurable and devastating disease. Although the error was tragic, the overall goal of the program was noble. Dr. Hughes was one of the few doctors capable of reading the DNA of a fertilized egg. His leaving the NIH was an unfortunate event, but his move into the private sector was a boon for couples like Brian and Hilary who were desperate to give birth to healthy babies.

The procedure Dr. Hughes had mastered was called "preimplantation genetic diagnosis," or PGD. Today PGD has become exceedingly commonplace for couples seeking IVF treatments. Still, a loud cry for caution continues to taint its perception by the public.

In his book *Far from the Tree*, *New York Times* reporter Andrew Solomon wonders about a world in which technology develops the capacity to weed out "disabilities" like deafness and Down syndrome. Many have argued that both deafness and Down's are more than just a "lack" of some perceived ability, and many people who were born with them have enriched our world. Similarly, Solomon wonders if we start with naming something an illness, how might we end up "defining" illness in general? Once we find the genetic link for a perceived "weakness" or for "stupidity," will we weed those babies out too? And where does it stop? Will we weed out eye colors? Homosexuality? Ethnicities? Will we end up with a homogenized world that no one living today would recognize?

Of course, Dr. Hughes would probably argue that saving a child from a foreshortened life and compromised functioning is hardly the same as practicing eugenics. If modern science can prevent people like Hilary and me from bearing children who suffer from a devastating illness, why would we choose not to avail ourselves of this possibility just because we're worried that it could mean that one day science would eliminate redheads? It *sounds* ridiculous.

Sometimes people who are uncomfortable with the idea of PGD ask me, "What if your father had never been born?" I understand their argument. What if in choosing to carry an embryo without illness, you are choosing against an embryo that would have been a great guy? I would argue that the person you end up creating might assert the opposite. But the fact that we live at a time where we have these choices, to my mind, is both humbling and empowering.

"We all throw genetic dice when we have children," I heard Dr. Hughes say on *60 Minutes*, "but when you know that those dice are loaded, that there's a really reasonable chance that your baby will have an incurable, dreadful condition, you're looking for an alternative."

That's exactly how Hilary and Brian saw things. They were thrilled to be able to send their fertilized eggs to Dr. Hughes after Dr. Kricket passed along the pertinent information to him, providing him with the map to our faulty gene. Then Dr. Hughes was able to choose two healthy blastocyst eggs. Those eggs were implanted in Hilary's uterus the next day.

Nine months later, Addison Rose and William David were born

healthy, beautiful, and free of the family gene. Both Hilary and Brian are proud of the way they handled the complicated fertility treatments that brought them their children.

Their son Billy was named after my father. Every time I think about it, I am overwhelmed by the miracle that allowed this sweet boy to have his grandfather's name . . . but not the gene that killed him.

TWENTY-THREE

hadn't thought much about having children myself. Before I moved to Prague, I took a class to become a certified English teacher. One day, as the class chatted before our lesson, I mentioned casually that I didn't want to have children. This group of people I had known for about ten days audibly groaned. One of them said, "You're twenty-three, you'll change your mind." Another added, "You are totally the kind of person who says that, then ends up having kids."

I didn't fight back too hard. I understood that saying you didn't want children was easy to do when your fertile years stretched out before you like a giant field of wildflowers. It's when that field begins to resemble a Mad Max set that you become pretty sure your life's purpose is motherhood.

As I got older, however, my feelings stayed largely the same. I liked children. I could sense that I would like having them. But having them was never my priority. In fact, if I could have just gotten knocked up without catastrophic genetic repercussions, I am positive I would have had children, or anticipated having them soon.

My move to New York, however, had reawakened something in

me that resembled hope. What therapy had failed to do, relocating near several good friends in a town that practically hands out business cards to dream jobs accomplished completely. I remember skipping up subway steps one day and emerging into a warm spring afternoon. I felt happy, even euphoric. I began writing. I stopped smoking. I wore short skirts with high boots that covered the swollen ankles I no longer bothered to think about.

I was living with one of my oldest and dearest friends, Lisa, a successful landscape architect. One night, we went out to a neighborhood bar to celebrate another friend's birthday. Outside, a tall blond man with broad shoulders was smoking a cigarette. I felt a pang of disappointment because, as a new nonsmoker, I understood the risks of attempting to date a smoker so soon. But once I was inside the bar, I couldn't resist following him when I saw him get up and join the bathroom line. He introduced himself as Aaron, and quickly relayed the fact that he was from Ohio—the Cleveland area.

Aaron's velvety voice was designed for radio, but he had never settled on a specific career. Instead, he aggressively committed to and then dropped whatever career path most interested him in the moment. Aaron had worked on iron-ore ships on the Great Lakes, he had gone to broadcasting school, and he had been the newsman at a country radio station in his hometown of Oberlin. He studied at the Cambridge Culinary Institute and had cooked in kitchens. Most recently, he had started a career in construction, doing craftsman work on some of New York's most historic residences, including the Dakota.

The night I met Aaron, Lisa met his friend Elijah. Lisa blew off Elijah's advances and Aaron asked me to go biking the next day. I said yes. The next morning I canceled, largely because I didn't have a bike, but also because I wasn't sure I wanted to go out with him.

The following week I changed my mind and texted Aaron to meet me after work for a drink. Our first date lasted from happy hour to almost one A.M., when he walked me to my subway train and gave me a hug good-bye.

By our third date, we had begun to uncover each other's quirks. For example, Aaron told me that he figured the moon landing was probably a hoax and he sometimes voted Republican. It was also the first time he saw my swollen ankles. The fact of my gene came up on that date, but Aaron didn't ask many questions. When I asked him what he thought when he first saw my ankles, he jokingly but truthfully deadpanned, "Well, it isn't a trait you look for, but it wasn't a deal breaker."

I was lucky enough that a man like Aaron came into my life when he did, given that I might have settled for pretty much anyone with health insurance who could disregard my complicated health history. Several months later, in January 2008, he moved in with me. Around the same time, Lisa moved in with Elijah five blocks away.

TWENTY-FOUR

Valerie was my dad's first cousin, my grandmother's oldest niece, and my aunt Norma's oldest daughter. I wasn't very close with her, at least in part because she was of my father's generation and she lived in New Jersey. I was closer with her son Jordan, my second cousin, who was around my age and lived nearby in Brooklyn.

On April 20, 2009, I heard through the family grapevine that as fifty-nine-year-old Valerie was driving home from the supermarket in her suburban New Jersey neighborhood, something had gone terribly wrong. She was on the phone with a friend when suddenly her vision began to dim, then grew fuzzy. She pulled over and told her friend she'd call her back. With great difficulty, she typed out the numbers on her cell phone to reach her husband, Michael.

"Something's wrong," she told him.

"What do you mean?" he asked.

"I think I'm having a stroke."

Valerie was starting to feel disoriented. She told Michael that she was dizzy, that it had been difficult to even dial his number.

Initially, he wasn't sure if she wasn't just being dramatic. But Val insisted she couldn't drive the car, so Michael, who was in his own car twenty minutes away, drove to where she had called him from. He was with their younger son, Storm, who had recently joined him in his pest-control company. Michael wasn't particularly worried, so he and Storm continued to discuss matters pertaining to their business as they had been doing before Val's call.

They followed Valerie's path from the supermarket toward their house until they saw her car lazily parked on the side of the road, hazard lights blinking. Michael knew the minute he saw his wife that something was very wrong. The entire left side of her face was drooping. As he walked her over to the passenger-side door of his car, he thought he noticed that one of her legs was floppy.

Michael drove to nearby Hackensack University Medical Center. Hackensack has one of the few bloodless units in the area. This is particularly important to the Jehovah's Witness community, which Valerie and Michael had joined before they got married. Devout Jehovah's Witnesses don't believe in receiving blood transfusions for any reason. As Michael pulled Valerie from the car, her legs suddenly buckled underneath her. He called out for help. Valerie was swiftly taken from her husband's grasp and rushed to the emergency room.

They were pushed through triage, and doctors quickly diagnosed Val as suffering from an "intracerebral hemorrhage due to vascular malformation."

Val was right. It was a stroke.

Fortunately for her, she didn't need a blood transfusion. Even more fortunate was the fact that Hackensack happened to have one of the leading trauma centers for stroke patients in the world. Michael sat with Storm in the waiting room for about half an hour. When a doctor eventually joined them, Michael asked him for an honest assessment. He didn't want anything sugarcoated.

"People in this condition have a fifty-fifty chance of survival," the doctor told him. "If she does survive, we don't know how much damage has been done." Suddenly Michael wished he had asked for a little sugarcoating.

Three weeks after he'd met his wife at her car and brought her to the hospital, the only positive response Val was giving the doctors was slightly curled toes when they swiped the bottoms of her feet. It was a good sign, Michael was assured. But everything remained touch and go. Michael felt like he was constantly shuttling back and forth between hopefulness and making funeral arrangements for his wife.

When Aaron and I came to see Val during those first harrowing days, the waiting room outside the ICU was packed with people from her church. When it was our turn to go back and see her, we went with Aunt Joanie.

"Her mother isn't alive to be here," Aunt Joanie told us, having come to the hospital almost every day since the stroke. "I have to be here for her mother."

We stood over Valerie, a cousin I only vaguely knew. Would she even recognize me if she could open her eyes? Her head was bandaged and her face was bruised. A tube in her mouth was breathing for her. I'd never seen her without lipstick.

Aunt Joanie rubbed her hands and told her she loved her. I kissed her cheek. Even Aaron, who'd never met her, touched her shoulder gently.

No one in the family, as far as we knew, had been stricken with a gene-correlated illness since Uncle Norman died in 2002. Everyone was living in the same kind of oblivious lull they had enjoyed during the thirty years between Uncle Nathan's death and my father's.

My cousin Valerie is my first cousin once removed. That means she is the first cousin of my father. Val and I are first cousins, separated by one generation. Valerie's sons, Jordan and Storm, are my second cousins. Dr. Kricket usually refers to Val as my aunt because she is of my parents' generation, although she is not a sibling of my parents. Val's mother was my great-aunt, and all the descendants of my great-aunt are my cousins, regardless of how our generations match up.

Valerie was more serious than her younger sister, Suzanne, and her big brother, Ken. Always put together in beautiful leather jackets and with impeccable makeup, Val was cool. She had countless friends, and her husband adored her. Her older son, Jordan, had an amazing artistic ability that his grandmother, Aunt Norma, had been immensely proud of. He grew up to become a jewelry designer and international gem trader. Her younger son, Storm, stayed closer to home and became a leading figure in her church. Valerie had two great kids, a steady marriage, and a religion she loved. She also had her mother's bad gene.

She hadn't been experiencing any problems because of it, at least to the best of our knowledge. Val's health issues were vastly different from what anyone else in the family had gone through, so there was very little reason to try to correlate them. At one point, as I'd overheard at Aunt Norma's funeral, Valerie had undergone a splenectomy to remove an enormous and grossly spongy spleen from her abdomen. No one else in the family, as far as I knew, had suffered from the same problem. But Val and the rest of us did share one important thing: no one knew what the hell was going on. And another thing: her platelet levels had once been very very low.

Why were they low? Why had Val's healthy spleen been engorged? Why was an otherwise healthy woman in her early forties having to undergo such a catastrophic procedure? And why in her late fifties was she now lying in an ICU on a ventilator?

Perhaps we had all grown so accustomed to hearing "I don't know" in lieu of answers that their absence actually seemed normal. But even Aunt Norma, a practicing nurse for forty years who lived through her daughter's first invasive procedure, never compared Valerie's "I don't knows" with her own.

As a woman with the family gene, it seemed like she would have an edge, at least compared to the men. Her mother lived into her seventies. My grandmother, Val's aunt, had made it into her eighties. Any whispers about my great-grandmother, Val's grandmother, dead by fifty-four, were disregarded. *Shh*, I told the voice. *That was the 1950s.* And anyway, Valerie had a *stroke*. How was a stroke related to lymphedema or massive ascites? It wasn't. Valerie was exhibiting no swelling; she and I didn't even share ankle problems. Meanwhile, Grandma Mae hadn't had a stroke. Aunt Norma hadn't had a stroke. Uncle Norman, my dad, Uncle Nathan . . . none of them had had a stroke.

Meanwhile, Ken, Valerie's brother, Aunt Norma's oldest child and only son, was told definitively that he didn't have the murmur, although he remained skeptical for years, having watched his cousin Norman receive the same vote of confidence before falling desperately ill. Ken's two daughters didn't have the murmur either, which seemed to squarely put him in the clear. Like my own father, Ken's girls would have inevitably inherited any gene on his X chromosome. If he'd had the gene, so would they. Suzanne, Aunt Norma's youngest, conversely, had the murmur and therefore the gene like her sister, Val. But she had no health problems to speak of.

When Valerie's two sons and Suzanne's daughter and son were tested, neither mother recalls ever asking for or hearing their results. It might have felt easier to carry on not knowing, and Dr. Kricket made it her business never to offer information to families without prompting. She knew if any of us was going to get sick, we'd get sick. But in the meantime, living confident, healthy lives was more important than miring us in a fear that could easily prove illusory.

So none of Valerie or Suzanne's kids knew if they had the gene, but at the time, knowing really would have mattered only to one of them: Suzanne's daughter, Denise, who was about to start a family. Though a year younger than me, Denise married when she was twenty-two and gave birth to her two children before she knew the risk that passing our gene posed to them. At the time when Val was convalescing from her stroke, Denise's children had still never been tested.

Valerie stayed in the ICU for two full months before she was transferred to a rehab facility. Her husband, sister, and sons remained vigilantly at her side. Val had so many regular visitors that the hospital staff began trying to figure out who she was. With all that traffic, she had to be pretty important.

Valerie's recovery was touch and go. There were setbacks and leaps forward and then stumbles throughout. Her husband, Michael, would say, "Valerie, if you can hear me, squeeze my hand." And she would. A little while later he'd say, "Lift up your knee." And she'd do it. But there was no facial reaction. No recognition. No Valerie.

Until, one day, there was. Slowly but surely, Valerie learned how to talk again, and walk again. She grew stronger and stronger until her doctors became confident that she might even enjoy a quality of life similar to the one she had known before the stroke. Michael credits the constant stimulation from friends, family, and church members for helping his wife ultimately regain much of her physical and mental function.

While Aaron and I were driving away from the Hackensack University Hospital ICU and a motionless Valerie days after her stroke,

I leaned my head on the window of our car. It felt good to be out of the hospital. It felt good to not have to go back.

"Do you think . . ." I started to say what I was thinking to Aaron. Was there a correlation between Val's stroke and my gene? Then *Nah*, I probably decided, swatting the idea away. *No way.*

I watched as the Bergen County landscape melted into New York City slurry, and then firmed up again.

TWENTY-FIVE

A year and one rescue dog later, Aaron proposed to me and gave me his grandmother's ring. He'd had it resized and fitted with my birthstone, a garnet, in the one setting where a diamond chip had fallen out.

I called my mother to tell her about our engagement. Her first impulse, even before congratulating me, was to suggest that I always leave my engagement ring on, even when I slept. I am prone to losing things. I am also prone to dropping them: in the bathroom especially, you can count on me to flick earrings, fumble lipstick, and certainly mishandle engagement rings. I have more than once referred to myself as "floor-blind," and almost always rely on Aaron to find anything I have dropped, or give up hope. I took my mom's advice and slept in my ring.

I was thirty-six years old that morning in early spring of 2010 when I woke up to my left ring finger puffed up from my bottom knuckle to about four inches above my wrist. I didn't know that day that I was only two years older than my great-grandmother Mae had been on the morning she woke up to swelling in her arm.

The choke hold my engagement ring had on that finger that morning was alarming, to say the least. Our dog zipped under the bed to cower. Aaron was immediately at my side. "Oh, my poor baby," he said calmly. He clicked his tongue and comforted me with well-timed "aw"s as he guided me from the bed to the bathroom. He left me momentarily clutching my hand under cold water until he returned with the bottle of Dawn from the kitchen sink. He poured most of the bottle onto my hand. As Aaron gently smoothed the soap-smothered ring over the firm gel of that chyle-filled knuckle, it finally slipped off.

My hand, while not wholly remarkable in appearance, now boasted a permanent oblong-shaped bubble beneath my ring finger. It relaxed into a pitted oval when pressed, then slowly re-formed as the fluid settled back into place. We stood there staring from it to each other as the whir of the fan motor lent a sound track to our Sunday morning.

"We'll call Dr. Seidman," Aaron intoned in his deep, soothing voice.

I leaned into him as he hugged me. The dog scampered out from where she was hiding under the bed to include herself in our embrace.

We didn't get around to calling Dr. Kricket because a few days after that harrowing morning, my mother fell and broke her hip. I flew to Columbus. When I landed, I realized something immediately perplexing: my foot was trapped in my shoe.

As soon as I was alone inside my mother's house, I rolled up my pant leg as far as I could. Mind you, these were not skinny jeans. The skin of my calf was pulled taut so that it had taken on the look of a loaf of seeded rye bread with flecks of dark stubble that had nowhere to hide. The foot itself, sockless, was puddling over the rim of my shoe. The skin itched a little bit, but other than the discomfort of the weight itself, there was no pain.

My ankles had been swollen for the last ten years. But this swelling in my leg was new and, on the heels of my swollen hand, alarming. Even stranger, the swelling in the ankle of my left foot had diminished substantially. In fact, that ankle, once the more swollen of the two, suddenly looked normal.

I was surprisingly unfazed. Some people would have driven straight to the nearest Urgent Care. I drove myself to a nearby Walmart to buy a replacement pair of shoes and find a pair of scissors to excavate my foot. Luckily, Walmart had scissors, shoes, and a pants aisle. I cut from the hem of my jeans leg and tore the fabric up to the knee, where the swelling began. I worked carefully, sawing at the leather, finally leveraging the scissors beneath my heel. The red Danskin clog slid off. I replaced it with an unlaced yellow work boot in a size bigger than what I usually wore. While fashion wasn't at the top of my must-have list just then, I felt like the shoes worked as a nineties throwback, even if I tripped a little over them when I tried to walk.

I changed into a pair of wide-leg black yoga pants and took my pile to the front of the store to pay.

"I'm wearing it all out." I shrugged to the salesgirl, handing over a shoe box with a pair of clogs inside—one mangled—tags from my new yoga pants, and an empty scissors wrapper. I caught her judgmental side eye, but opted to ignore it.

I didn't go to the doctor. I didn't even call Dr. Kricket right away. My now-swollen leg and hand felt like an extension of my swollen ankles, which I had already endured for five years. I knew that women in my family had swelling in their limbs, so it didn't scare me so much as disappoint me. My swollen leg suddenly meant I couldn't wear skirts, and almost all of my shoes, barring my new extra-large work boots, were out of the question on my right foot. No part of me worried that it was anything more than cosmetic. Anyway, if I *was* about to be ravaged by lymphedema, then it was probably better to deal with my mother first.

Later that night, I showed my leg to my sister. She and I didn't say much on the subject.

"Does it hurt?" she asked.

"No," I answered.

"Grandma's legs look like that," she added.

"I should get some of those fitted stockings she wears," I answered. She agreed. And that was all we really said.

A few weeks later, a college friend came to town, prompting a group of us to meet up in midtown Manhattan for pizza. It was a hot spring night and we drank several pitchers of beer. As I made my way home on the subway, I could feel my leg pulsing against the fabric of my jeans. Already relegated to wide-legged fashion, and just barely figuring out what shoes my foot would fit into, I worried that an expansion of the condition of my leg might push me over the edge in several ways. I called Dr. Kricket and made immediate plans to visit.

TWENTY-SIX

On June 15, 2011, I visited the Seidman lab to present my newly enormously swollen right leg. Dr. Kricket's very first question was pretty basic: "So, do you have a doctor yet?" The answer was the same as it had been since college. "Aren't *you* my doctor?"

Dr. Kricket is a researcher, not a diagnostician. While she was more than happy to weigh in on my medical predicament, she simply wasn't in a position to serve as the leader of my health and wellness team. I told her that Aaron's boss understood that I would go on Aaron's insurance as soon as the ink on our marriage license was dry. But right then, she was all I had. My swollen leg. My swollen hand. She was all I had.

She ordered an MRI under the umbrella of her study. As I lay still inside the narrow tube, hearing the beeping, clicks, and squeals of a machine that, incidentally, I still can't believe is the best we have to offer in a post-man-landing-on-the-moon world, I truly believed that Dr. Kricket wouldn't be able to tell me much.

Joselin, we don't know what caused your leg to swell. We don't know why you will need to learn phrases like "removable orthotics" when you

*buy footwear so that you can add an inch of depth to your shoes by remov-
ing the insert. We don't know why you can't wear your engagement ring
anymore.* Shrug. *You have a bad gene.*

After the MRI, I sat alone in a deserted waiting room. When
Dr. Kricket came in, we were the only two bodies within that quiet
space. Rows of empty chairs seemed to hold the apparitions of the
people who had previously filled them.

Dr. Kricket sat next to me, and then she did something surprising. She took my hand.

"I need you to get a doctor," she said, leaning in, her blue eyes
more intense than usual.

"Okay," I answered, bubbly, completely unwavering in my expec-
tation. "Next month."

"Meanwhile, dear," she went on, "I need you to be careful you
don't throw up."

I was about to laugh when suddenly it occurred to me that she
wasn't making a joke. She was telling me something. Something
about my health, and it wasn't the status quo.

"You have a blocked portal vein in your liver. Now, you seem fine
and I'm not worried"—her sentences were quick and staccato—"but
your spleen is big, which explains your low platelet count."

In the recesses of my mind, I thought about someone else who
had a blocked vein and a large spleen. "Mushy," did they call it?

"Like Valerie?" I asked.

"Yes, exactly like Valerie."

"Was her portal vein blocked?" I asked.

Dr. Kricket nodded. "But there's a treatment," she added. "A procedure to reopen it. Val had it and it worked."

My dad's portal vein had been blocked too. His treatment had
failed.

Why hadn't I known about Val's portal vein? Wait. Had I known?
I was suffering from momentary amnesia.

"Now, I want to tell you again, you do not have what your father had, and I know that you are okay, just like your grandmother."

"But I shouldn't throw up?" I tacked on at the end. Why? Why shouldn't I throw up?

"Right. No throwing up."

"And why does this explain my platelet count?" There were too many questions and it felt like I was asking none of the right ones.

"Your large spleen sucks up the platelets," she was saying, but the room had grown a bit echoey and I wasn't sure I could process what she was saying. First of all, how the hell do you not throw up if you have to throw up? Second, wait, no, *first* of all . . . Why? Why on earth should I not throw up? I do not enjoy throwing up. I do not ever want to throw up. But that isn't to say that at various points in my life I haven't, against my will, thrown up.

I led with this concern. "How do you not throw up?"

"Have your doctor prescribe you this . . ." and she wrote down the name of the antinausea pill.

If Dr. Kricket specifically answered my question about why I couldn't vomit, I can't remember what she said. Similarly, the note about my platelets getting sucked up into my enlarged spleen also got sucked out of my brain by something else only minutes later.

My friend Molly picked me up from the Brigham, where I'd met with Dr. Kricket. Molly is a nurse. She was Aaron's best friend's girlfriend, and when I met her, we had fallen in deep, unabashed friendlove. Having gotten ordained online, Molly had become the go-to wedding officiant among her friends. She was going to officiate at our wedding next month. Molly had the kind of temperament that was also perfect for a nurse, even if she had a tongue piercing, ear gauges, and tattoos all over her arms and back. She was gentle but decisive, confident but empathetic. She had been working for a year at one of New York City's largest burn units. When I told her I had to come to Boston to see Dr. Kricket, she told me she'd drive me. She had spent the afternoon with friends, but was waiting to pick me up when I emerged from the hospital at the end of the day.

About ten minutes later, I explained that there was something about a blocked portal vein in my liver and an order not to throw up.

"Why is that?" I asked her pointedly. "Why did Dr. Kricket tell me not to throw up?"

Molly directed a brief side eye my way, then squared up her shoulders, both hands on the steering wheel. "You know," she began, "I've only really heard of it in old alcoholic homeless people."

She explained that when someone is in liver failure, for example an alcoholic, the portal vein, which is a main pathway through the liver, the thing that Dr. Kricket had said was blocked in me, sometimes became blocked.

The vein fails, or clots with blood. Because the blood still has to find a way to pass through, alternate channels open up. They call these "varices," like varicose veins. Imagine that a car accident blocks a five-lane highway through your town at rush hour. The cars are routed onto the smaller streets around it. Now, let's say the city decides to expand the side streets to accommodate all the cars that usually take the five-lane highway. Instead of adding cement, they just spread out the cement already on the road and thin it out to fit the traffic. In a similar way these smaller channels expand to take on their new role as the primary supplier of blood to the liver. These expanded side streets typically run all throughout the digestive tract, through the stomach, and ultimately up into the esophagus. They are more delicate, holding more blood in unnaturally expanded, weaker channels. It turns out, Molly explained, it's the ones in the esophagus that we need to worry about.

"Why?" I asked, and let me just say that even after everything I'd heard up to this point in the day, I still wasn't scared.

Molly side-eyed me again. "Well, if an esophageal varice bursts," my friend explained, "you could bleed out your entire blood supply through your mouth."

You. Could. Bleed. Out. Your. Entire. Blood. Supply. Through.
Your. Mouth.

And with that, I stopped sleeping.

I got married forty-five days after seeing Dr. Kricket.

We were doing it in western Massachusetts, near Pittsfield,
where both of my parents had grown up. We had chosen a beautiful ski resort called Jiminy Peak where we had stayed several times,
always in the summer. Aaron and I were both from Ohio, where ski
club is really something school kids just sign up for so they can make
out on the bus. Mountains are in limited supply in Ohio, but this
New England ski resort in the summertime had become someplace
we liked to go. The resort had a small eco-friendly amusement park
that included a gravity roller coaster that barreled through the trees.
There was also a heated pool and a twenty-person hot tub. They had
an alpine slide, and Aaron and I had joked about riding down it
while holding hands as a dramatic entrance to the ceremony.

I still didn't have a doctor, so instead I carried a bottle of antinausea pills in my pocket and a low-grade near-constant dread in
my heart. On a lighter note—or perhaps a heavier one, given that I
was about to get married—I had grown accustomed to my bloated
leg, but not yet to the fact that my wedding shoes no longer fit my
right foot. The shoes I had purchased six months earlier, I had loved.
They were comfortable off-white ankle boots.

But the day of the wedding, when Aaron and I awoke, I still
didn't have a replacement plan beyond my yellow Walmart work
boots. I had brought the ankle boots, and as I fought to stuff an extra
two-inch girth of lymphedema into them, I had started swearing to
myself.

As I sat on the floor trying to wedge my right foot into the shoe,
Aaron grabbed my hand and shepherded me into the car. I'm pretty

sure I groaned, rolled my eyes, and stomped my foot en route. We drove past several groups of our guests. I felt surges of anger. Amy had flown in from Prague. Another best friend from high school I rarely saw had come in from California. Our wedding was a destination for almost everybody. I had been drawn to the idea of keeping my guests contained for three whole days of celebration. I deserved it. I had a blocked portal vein.

"We'll be quick," my almost-husband promised me, squeezing my arm.

Aaron guided me into a nearby mall and over to the nearest bargain shoe outlet. I became parallel-universe Cinderella as my prince Velcro'd a pair of hideous seven-dollar beige house slippers around my bulbous foot. It fit! It didn't have a choice. It was Velcro. Aaron bought the shoes for me and drove us back to get ready.

Later, as we walked down the aisle together at our wedding, my long dress covered my feet. My smile covered my internal vascular highway, filled with collateral, presumably poorly paved pathways. I smiled at Valerie, who had come with her husband, Michael. A year out from her stroke, she was wearing lipstick again.

In front of everyone we loved, I read my written vows, which were characteristically long-winded. My husband began his by announcing, "I will be delivering the Gettysburg Address to Joselin's Declaration of Independence." Everyone laughed, and I was grateful to him for knowing how to keep things light.

Molly pronounced us married, which was our cue to pile onto the ski lift, our chair bearing a "Just Married" sign amid dangling cans. Our guests followed in chairs behind us for a champagne toast by my new father-in-law at the top of the summer-green mountain. My mother remained wheelchair bound, having faced countless setbacks in her recovery from a hip fracture. She didn't join us at the top of the mountain but she had made it through the ceremony, and would even finally make a speech at the reception later that night. The day was perfect with sunshine flooding the valley. My slippers dangled beneath me as the sloping Berkshire Hills met us on our way up to a green summit, then fell swiftly away on our way back down.

Later, only one person, a cousin of Aaron's who knew nothing about our gene, and who thankfully hadn't noticed my swollen leg, pointed at my footwear.

"Smart choice," she commended, and we high-fived.

TWENTY-SEVEN

Not every genetic illness is inherited. Conditions like dwarfism have multiple genetic lines, each one starting with a different founder—the individual with the original genetic variation. For example, the gene for dwarfism is a relatively common mutation. One might say it's a delicate gene. Whether or not it mutates is entirely random. Once a person with that genetic variant has a child, he or she might pass the gene for dwarfism to his or her offspring.

Hunting down founders has become easier since the advent of the Internet. During my father's illness, the technology was still being developed. But today, when it's used correctly, answers can be uncovered in a matter of days.

Bo Bigelow, my friend's cousin, has a daughter named Tess, struggling with global developmental delays (the six-and-a-half-year-old had the developmental equivalent of a one-year-old) and a form of blindness related to cortical visual impairment. Her eyes worked, but the connection between those organs and her brain made interpreting the world around her complicated. None of Tess's doctors or specialists could figure out what was causing her problems.

Finally, the doctors found a genetic variant in Tess that her parents didn't have. The doctors wished the Bigelows luck and sent them on their way. Tess was, as far as they knew, "patient zero" with this condition. Not knowing what that variation meant for their child (would she live another year . . . another ten?), the Bigelows went home.

Bo hated it. He didn't want to just watch his daughter and wait. He wanted answers. First, he went online and continued the six-year Google search he'd started when Tess first began showing symptoms. Then he contacted a family he read about who had done a wide-reaching search for answers to similar health issues in their own child. He was instructed to build a web page and focus on making it as accessible as possible to anyone looking for answers. Instead of doing a Google search, he would become the Google result.

That's how I first heard about Tess. My friend forwarded me Bo's website. The way Tess's information had spread was—forgive the comparison—practically genetic, first through Bo's family and then into the greater population. A friend of another of Bo's cousins uploaded the information to reddit, where a scientist at Baylor University saw it and passed it on to a colleague.

The genetics department at Baylor had already brought together eight cases similar to Tess's, all with a similar genetic variant. Tess is the youngest. The oldest is thirteen. But the good news is that all the children seem healthy aside from their disabilities. None of the children in the study are biologically related, which means that their genetic variation is likely to have randomly happened in the sperm or egg that made them, or was a spontaneous mutation that occurred when they were in the womb.

Spontaneous mutations cause what is called a "mosaicism." My great-great-grandmother Ester Bloom, whose ankles, according to family lore, swelled while she was hanging laundry on a Brooklyn rooftop, might have been the origin of our family gene. If so, it's possible only some of Ester's cells carried the genetic variant or mutation. Once the sperm entered the egg, it remained one cell for around thirty hours, and then it divided into two cells. If the mutation or variant happened during this first split, only 50 percent of Ester's cells carried the gene. If it happened fifteen hours later, during the next split, only one in four of her cells carried it, and so on. We will never know for sure. Our only clue that Ester shared our genes are those swollen ankles staring out at us from photographs. But if we're right, and the information for this bad gene that may have existed only in *some* of her cells passed to her daughter, Ester Bloom is the founder—or patient zero—of our gene.

During the summer of 2016, the first Founder Population Summit met in Haifa, Israel. The action plan for the summit was simple: look back over the Ashkenazi Jewish gene pool and figure out where things had gone so genetically wrong. The Ashkenazi gene pool has a disproportionately high number of genetic mutations, largely because of geographic and political isolation as well as a religious standard placed on marrying other members of the population. The Ashkenazi gene pool owes its lot to forced homogenization— sometimes self-imposed, but also thanks to anti-Semitism and prevailing political ideologies.

Some of the genetic variants seem to benefit the population, like a gene that is currently being hunted down and is affectionately known as "the Super Bubbe gene." The Super Bubbe gene is believed to help the mostly women who have markers for illnesses like cancer and Alzheimer's live deep into old age. Other genes are harmful, like those for Tay-Sachs disease and Bloom syndrome, both of which usually kill their young sufferers before they reach puberty. There are at least eleven deadly genetic conditions for which one in four Ashkenazi Jews is believed to carry the variant. When two Ashkenazi Jews prepare to have children, they are screened for all of them.

Aaron decided he wanted to try out a new service, where you send away your DNA and find out your ethnic heritage. Just how homogenous was my DNA? We used a boutique company called 23andMe. They billed themselves as "the largest DNA ancestry service in the world," offering basic genetic information to anyone who wants it for the low, low price of $99—at least, they did when we tried it in 2014. Regardless, when you consider that in 2007 a single sequenced genome cost $350,000, you have some perspective on just how far we've come.

For now, 23andMe only allows users to see the places from which their genes originate with limited information about their health. It delivers a circle graph broken into pie wedges letting you know what percentage Scandinavian you are, if you are any percentage Scandinavian. It also connects people who share your genes—distant cousins and the like. It's supposed to be fun.

Much of my adult life had been spent focused on my genes. I wasn't sure sending my DNA to 23andMe would be as much fun as Aaron thought it would be. But I supposed it wouldn't be *un*fun, so I agreed.

The tests arrived in a small white box a few days later. We spit into a small plastic container, sealed it up, and sent it back. Ten days after that, our results appeared on a website. Aaron's revealed that he was 40.7 percent Irish and English, 13 percent German/French. There were additional bits and pieces from other places, mostly in Northern Europe.

My results were a little different. My circle, which was supposed to be divided into a nice little assortment of pie wedges, was just one giant orange circle. Ninety-eight percent Ashkenazi Jewish ancestry. Ninety-eight percent. I have to admit I was surprised. I figured someone somewhere along the way might have had a fling with the milkman. But no. The other 2 percent of my circle had something to do with Neanderthal genetic code, which is a little joke 23andMe mixes into everyone's results.

One of my best friends, also of Ashkenazi Jewish descent, recently underwent genetic testing in the midst of fertility treatments. Her DNA was riddled with flagged genetic markers. She was even told she carried one gene that, if paired correctly, would yield a hermaphroditic child. Her husband, a Maine-born lumberjack type of mixed Northern European ancestry, was also tested. The doctors couldn't believe their eyes. His genetic sequence, they told him, was that of "a superior specimen," a funny anecdote they both—but especially *he*—still like to bring up at dinner parties.

Martin Fugate made his way to Kentucky in 1820 to lay claim to a land grant. An orphan, Fugate had grown up in French workhouses. For all intents and purposes, he was healthy and attractive enough to find himself a red-haired American wife with pale white skin. He and his bride, Sarah, settled on the banks of a place called Troublesome Creek. But there was something causing Martin Fugate trouble. Martin Fugate was *blue*.

Although historically we have used words like "red," "yellow," "white," and "black" to describe skin color, from kindergarten onward most of us understand that we're talking about a subtle variation in a skin *tone* and not the actual color of the skin itself. But Martin Fugate wasn't a "winter," with blue undertones to an otherwise peachy glow. He was blue, somewhere between "periwinkle" and "Smurf."

Now, as the generations wore on, whispers about pockets of blue people had spread throughout the mountain communities. Then, in 1960, a blue person oddly but somehow appropriately named Luna Fugate arrived at a clinic in Hazard, Kentucky, asking for a blood test. A sixty-seven-year-old nurse named Ruth Pendergrass took one look at Luna and expected her to die of a heart attack at any minute. She described Luna as "dark blue" with indigo fingernails.

Meanwhile, around the same time, Dr. Charles Behlen II was working an afternoon shift in the emergency room of a hospital at the University of Kentucky when a man named Luke Combs came in with his ailing wife. From minute one, Dr. Behlen couldn't have cared less about what ailed Combs's wife.

"Luke was just about as blue as Lake Louise on a cool summer day," he poetically explained to Bryce Nelson of the *L.A. Times* in an article published on November 6, 1974.

Soon Nurse Pendergrass and a hematologist named Dr. Madison Cawein found themselves trudging on foot through the mountains looking for the people the locals called "the Blue People of Troublesome Creek." Eventually, they found a brother and sister named Patrick and Rachel Ritchie.

Dr. Cawein is reported to have said, "They were bluer than hell!"

First, he and Nurse Pendergrass sat down and wrote out their family tree. It quickly became clear that the blue skin of the Ritchie siblings had been passed through their genes. Due to their relative geographical seclusion, along with a lack of adequate roads, blue people had been cropping up in pockets all over the region since the arrival of Martin Fugate. Cawein noticed that four families—the Fugates, the Combses, the Ritchies, and the Stacys—had, over six generations, frequently intermarried, most typically among cousins, but also, in one instance, an aunt and a nephew.

Dr. Cawein and Nurse Pendergrass drew the Ritchie siblings' blood. Upon doing so, they discovered that they both lacked a specific enzyme, diaphorase. Their findings revealed a rare condition called "methemoglobinemia," or met-H. Specifically, diaphorase repairs the protein molecule that carries oxygen in the blood. Without it, the damaged protein carries less oxygen, turning it blue, finally turning the skin of the sufferer blue. Otherwise, its impact is virtually nonexistent. Luna Fugate lived to be eighty-four years old and gave birth to thirteen not-blue children. People with met-H live healthy lives, even if they are not always so happy.

You've heard that it's not easy being green, but being blue is no picnic either. The clinicians who studied Rachel and Patrick Ritchie described two people visibly ashamed of their skin hunched over and avoiding eye contact with their doctors. Luna's own husband, who outlived her by twenty years, happily talked about many of his wife's blue-skinned kin, but shied away from admitting that his own wife was blue, even if everyone who ever knew her was quick to state otherwise.

Luckily for Rachel and Patrick, they learned that met-H can be treated with a daily injection of methylene blue—a literally blue substance that restores iron to the blood, turning it red.

Today, almost all of the blue people of Troublesome Creek have passed away. Benjy Stacy, forty-five years old, has skin that turns blue when he's cold or under stress. But the mutation that leads to met-H is recessive and exceedingly rare. Both parents must carry the mutation that causes it.

Small gene pools have been shown to lead to an increase of rare genetic conditions, largely thanks to the heightened possibility of two parents passing along the same gene. The fact that for every single gene in our DNA, there is a second gene coded to do the very same thing—one from each parent—actually makes us stronger. Our bodies, in a "survival of the fittest" spirit, often opt for the stronger, more successful of the two genes, also called "alleles." Richard Dawkins, in his book *The Selfish Gene*, suggests that alleles are in fact competitors—"rivals," he calls them—for the same job. When those two genes from each parent are exactly the same, by default the body must choose that gene, for better or worse, whether or not it contains less good or even bad genetic information.

Perhaps the most remarkable story about the detrimental impact of "inbreeding" (a mostly negative word indicating that two mates are closely related over multiple generations) concerns the Spanish monarchy. The Hapsburg line famously died out when King Charles II passed on without an heir as a result of his well-documented infertility. Severely mentally challenged, the thirty-eight-year-old king likely suffered from a multitude of recessive genetic conditions and a weakened immune system resulting from rampant intermarriage between siblings, cousins, and other family members throughout the Hapsburg line, a common occurrence among aristocrats of the time. King Charles's father was his mother's uncle.

Today, mores along with technological advances like cars, airplanes, and even the Internet impel and allow us to seek mates from a much wider variety of gene pools than ever before.

In 2013, an instance of an incestuous family showed the horrible toll of inbreeding (as well as pedophilia) in stark relief. Given the pseudonym "the Colt family," four generations of close family members were discovered living together in the Australian wilderness. Incest among brothers and sisters, parents and children, and even an instance of a grandparent and a grandchild—had led to catastrophic illness, mental impairment, and even a rare genetic condition that caused the death of a three-month-old. A teacher at the local primary school was reported to have overheard one of the Colt children say to another child in the playground, "My sister is pregnant and we don't know which of my brothers is the father." The statement led to the discovery of these dozens of family members of varying ages living together on an impoverished commune. The family had been so isolated that several teenagers did not speak properly, and none of the children had any practical understanding of hygiene. The repeated occurrence of mental disorders and physical deformities among the children seemed in equal parts due to a lack of proper treatment and to a multitude of rare, recessive genetic conditions caused by incestuous relationships.

Cases like the Colt family, as well as extensive studies on inbreeding in other species, offer fairly damning evidence that it is to our benefit that we mate outside of our familial gene pool. However, historically, we haven't always known this to be the case. Marriage between cousins was the rule far more often than the exception even in this country less than two hundred years ago. Even today, religious and cultural groups make a point of choosing mates from within their ranks.

Some statistics show that, worldwide, people are most likely to marry someone who lives within a twenty-mile radius of their home—or the length of a bike ride. But with the ubiquity of cars and airplanes, once-segregated groups have access to gene pools from great distances away. Today the Ashkenazi gene pool, like so many others, enjoys far more diversity. However, for people who have given birth to sick children, perhaps even for my family, the damage was already done. But it can be undone.

Benjy Stacy is believed to be the last blue person in his family. Met-H is a recessive condition, which means that in order to pass the gene, Benjy's partner would also have to have the gene. It is possible that after Benjy, the recessive trait for blue skin will just silently exist among his ancestors, unable to express, unable to turn anyone else's skin blue. Until one day when two carriers of the gene for met-H, now possibly distant relatives, mate. Then once again a blue person will be born.

The other possibility is that Benjy Stacy won't pass his blue gene. Neither will any of his relatives, both close and distant. In that case, the gene for blue skin will simply die out as naturally as it came to be.

There remain, of course, groups of people who prize religious or ethnic homogeneity, and I understand that there can be a richness to maintaining and cultivating a shared history and customs. However, I have grown skeptical about it being in our best interests to breed like with like. Genetically, there is some stark evidence that suggests we're stronger and better off when we mix it up.

TWENTY-EIGHT

I became fascinated with the BRCA1 variant and Angelina Jolie when her first article appeared in the *New York Times*, on May 14, 2013, largely because I had just spent a year mulling over my own preventive treatment. Preventive treatments usually come up when healthy people want to stay healthy and a genetic variant indicates that they might not.

Angelina's mother was diagnosed in her early forties with a deadly ovarian cancer caused by a gene variant called the BRCA1, commonly pronounced "BRAH-kah." The BRCA gene comes in two varieties, 1 and 2, and likely more will be flagged before this book goes to print. Both BRCA1 and 2 are actually stabilizing forces in genetic material. When they contain variations like Angelina's, however, they destabilize cells, which can lead to cancer. The information in Angelina's DNA from the BRCA1 is repeated in every cell of her body, over and over, in upward of thirty-seven trillion BRCA1 genes.

The name BRCA itself simply combines the first two letters of the words "breast" and "cancer." BR-CA. But it doesn't just *mean* "breast cancer." BRCA also elevates the risk for cancers of the ovaries and fallopian tubes. It also leads to greater risk of peritoneal cancer, which is cancer in the lining of the abdomen and the lining surrounding abdominal organs. In men, the BRCA genes increase the chance of prostate cancer.

After her mother's death at the age of fifty-six, Angelina became very public in her own battle with the BRCA1 gene, which she learned she had inherited from her mother, just as I had inherited my own bad gene from my father. It made me feel close to her. You know, me and Ange.

There are very few treatments for genetic conditions, like chest percussions for cystic-fibrosis patients and tapping for my family's illness. Today, prophylactic amputation, a procedure popularized by Angelina herself, has become a way to treat a possible, impending deadly illness. In May of 2013, Angelina Jolie had her healthy breasts removed. On March 24, 2015, she followed it up by having her healthy ovaries removed. She wrote about both in op-eds for the *Times*. If the cancer is most likely to form in her breasts or ovaries, she wrote, the simplest solution would be to remove the susceptible body parts. What many fail to realize is that although the chances of developing cancer might decrease by so many percentage points, you risk suffering from the side effects of these surgeries. It's a tough choice. Since these treatments are new, we lack long-term understanding of their impact.

In the summer of 2011, I was married and finally on Aaron's health insurance policy, so I took myself to a doctor who specialized in gastrointestinal conditions to examine the poppable varices in my digestive tract. I presented him with my MRI from the Seidman lab study, and he promptly passed me on to gastroenterologist Dr. Samuel Sigal at NYU, with a wry, "Wow, he's gonna love you . . ."

I brought Dr. Sigal, a renowned liver specialist and one of the top liver-transplant surgeons in the world, my father's medical chart and a story about my cousin Valerie. Unfortunately, the illness I was trying to describe was several generations along from patient zero and there were only fourteen people who had contracted it. This makes it difficult to explain to even the smartest people in the medical community. By all accounts, our gene continued to disregard our expectations with the curveball it threw us by involving our livers and its seeming disregard for our two X chromosomes.

Dr. Sigal's nurse practitioner was quick to tell me that reading my MRI made it almost impossible to believe that I could look so healthy sitting here in his office.

"You should look like you're dying," he said, with a big smile. "I mean, if I was only looking at your MRI, I'd say you needed a liver transplant, ASAP."

I don't think I replied. "Thanks" just didn't feel appropriate.

I knew that I wanted to stay a few steps ahead of the game. I took on the quest for answers with gusto. Dr. Sigal felt that I should consider a treatment called a "transjugular intrahepatic portosystemic shunt," or TIPS. Both my father and my cousin Valerie had undergone TIPS procedures. Val's was by all accounts successful. My father's had been a failure.

Dr. Kricket also agreed that a TIPS was something to consider. It was a minimally invasive procedure in which a radiologist tried to prop open my almost-nonexistent portal vein. The biggest complication—a swelling in the brain called "encephalopathy"—could be caused by a buildup of toxins in the blood. A shunted portal vein might also lead to a subtle but chronic state of confusion, or as Dr. Sigal put it, I could "lose my edge." I was gently reassured that this probably wouldn't happen to me because my liver seemed otherwise healthy.

Out of the gate, I liked Dr. Sigal and the radiologist he worked with, Dr. Hearns Charles, so much, so quickly, that I felt confident about putting myself in their hands and trusting their expertise. But many hospitals in the New York City system are overwhelmed with patients. My doctors took special interest in my case, thanks to its rarity, but they needed bite-sized answers they could tackle in the simplest way possible. I was far from simple. At that point, exactly two people worldwide had had a liver that looked like mine and shared my gene: my father and my cousin Valerie. Val underwent a successful TIPS procedure before her "mushy spleen" removal. With her portal vein open, the gastric varices throughout her digestive tract shrank. In my father's case, the shunt had simply clotted with stagnant blood and closed. So, based on the limited experience our family gene had with this procedure, the success rate was 50 percent. However, when considering statistics, I also had to factor in that ten years after her TIPS, Val had had a catastrophic stroke. Was her TIPS related to this? Did the 50 percent success rate come with an enormous, unattractive caveat? There was no reason to think it did, but given how little we knew, there was also no reason to think it didn't.

If shunting my blocked portal vein meant that I could lose my edge, not shunting it meant I wouldn't. I felt healthy and well. Attempting a procedure that could make me sick seemed counterintuitive. But just like a young Angelina Jolie responded to the possibility of dying of breast or ovarian cancer, I was motivated by the possibility of bleeding out my blood supply through my mouth. For now, though, as Dr. Charles, my trusty radiologist, put it, we'd start with a few simple tests.

———

I was awake as three arguably attractive men stood over me talking about whether or not they should shave my groin. Just the right side.

I was being prepped for a liver biopsy. This was my second test in two months. Dr. Sigal had already performed an endoscopy, in which—while I was under anesthesia—a little camera traveled down

my throat and into my stomach looking for varices along my digestive tract. The results contained the phrase "nothing too alarming,"
which is a mixed bag when you are looking at something that is one
puke away from a scene out of *Alien*.

The doctors decided to forgo the shave and forge ahead through
my jugular vein, in my neck. A wire threaded through my heart and
into my liver would allow Dr. Charles to grab a tiny piece of liver for
testing. I told Dr. Charles that I was under deadline and needed to
work later that day, so I didn't want to be medicated. He laughed and
gave me one of those "okays" that really mean "you're crazy."

There was no deadline, just an overwhelming need to feel every
minute of this. To imagine that I was doing it for my father, my sister,
my grandmother, and all of my cousins. As I lay on the table, wide-
awake, a local anesthesia was applied to the point of entry. I stared at
the machines beeping around me. I felt the wire as it snaked through
my chest. The bright lights hurt my eyes, which refused to close.

My bravery immediately dissolved. "Dr. Charles!" I called very
softly and carefully. A nearby nurse leaned down to look at me.

"Do you want to pick the music?" she asked.

The distraction worked. I forgot to say "give me drugs." Instead,
I found myself experiencing a moment of social awkwardness. Did I
want to pick the music? I hadn't noticed the music. But, sure. Except
my brain was blank. What tunes best fit the mood of the moment?
What was at once uplifting, calming, and also not entirely uncool?

"R.E.M.," I said. It surprised all of us.

Now, with what felt like a clothes hanger between my shoulder
blades I could hear Dr. Charles mumbling, then calling out, "Can
you page someone to . . . help me over here with . . . here, take
this . . . I need someone to . . ."

I closed my eyes.

*"Buy the sky and sell the sky, and tell the sky, and tell the sky fall on
me . . ."* R.E.M. sang on. *"This one goes out to the one I love. This one goes
out to the one I've left behind . . ."*

Shut up, R.E.M.

Dr. Charles pulled out the wire.

My liver, he told me, was, and I quote, "gorgeous."

"What this means," he went on, "is that I don't know." He wanted me to have another procedure, called an "angiogram," which checks blood pressure in my liver. For that one, they would definitely have to shave my groin.

TWENTY-NINE

Although it's redundant to say it, my blood tests continued to come back clean, and my liver was gorgeous, but everyone was still confused. I had begun taking a whole host of new pills designed to lower the blood pressure in my abdomen. I went off my birth control after a discussion with my gynecologist, who agreed that my fragile blood vessels didn't need to be further compromised by hormones, although it was clear she didn't entirely understand what the hell I was talking about and was more likely nodding blankly.

Aaron and I had decided to hold off on the conversation about children for the time being. We were told that blood supply during pregnancy increases by half, and my veins were already very delicate. I might be able to carry a child to term, if we decided that's what we wanted, once we had a better understanding of my physiology. Meanwhile, we bought condoms.

My third test, the angiogram, was slightly more invasive than the first two; during it, Dr. Charles inflated a balloon in my liver

in order to test my abdominal blood pressures. Three months after the anesthesia-free liver biopsy, Aaron accompanied me to
the hospital. I immediately asked for drugs. All the drugs. Every
drug. I did not want to hear R.E.M. and I didn't want to feel
wires. Before the procedure, I had to pee into a cup for a pregnancy test.

I couldn't pee. I sat in the bathroom for over an hour, feeling
like my bladder was a dud. Finally, I appealed to the nursing staff,
telling them I didn't believe I could be pregnant. I used protection
religiously. They took pity on me and let me out of the bathroom.

Two important things were true that day: (1) The blood pressures
in my liver, according to Dr. Charles, "didn't make any sense." And
(2) I was pregnant.

The amount of radiation that Dr. Charles used during the angiogram was not something you want to subject a human embryo to
unless you want it to incur superpowers. So at that point, there was
very little chance that the one inside me had survived. I went by myself to the high-risk ob-gyn appointment Dr. Sigal recommended,
because I was so certain there was no possibility of a viable embryo
that I didn't think Aaron needed to come.

At the ob-gyn, however, I learned that the embryo had made it,
although we wouldn't know for sure whether or not it would be able
to shoot spiderwebs from its wrists until its genes were tested. There
were other concerns. If I carried the fetus to term, there was a 25
percent chance I would die from a bleed out caused by one or more
of the collateral pathways from the blockage in my liver. What was
worse, there was an even greater risk that I would have to deliver
early in order to save me or the baby. Possibly very early. An early
delivery might mean a life of illness for our unborn child.

And yet I knew right then that I wanted to keep it. I was thirty-seven years old. This fetus had survived a microwaving. I suddenly realized I was *supposed* to have this baby. The doctor told me about a procedure whereby at the tenth week, I could have the genome of my fetus read to make sure it didn't have my bad gene and hadn't been fried into an evil leprechaun.

This procedure, called "chorionic villus sampling," or CVS, was briefly the go-to for genetic testing in the unborn. Before that, there was the slightly riskier amniocentesis, where amniotic fluid is extracted from the placenta for study. Today, a simple blood test called a "first trimester screen" doesn't go anywhere near the fetus. It simply uses a mother's blood for evaluation. In 2012, however, the first trimester screen wasn't being widely used. Looking forward to the CVS, I was certain, if this baby didn't have the gene, I was going to have it regardless of my own physical impairments.

Aaron was less convinced. I could see his point. I'd already leveled him with the admission that carrying meant I had a 25 percent chance of dying, that I could deliver early and have a very sick child, that there was nothing that wasn't dangerous about my plan. He agreed to wait at least until after the CVS to argue it with me. I scheduled my appointment for the earliest possible date.

The day before the CVS, on October 29, 2012, Hurricane Sandy hit. NYU, where my CVS was scheduled, was flooded, with twelve feet of water filling its elevator shafts. The entire hospital was thrown into chaos. I couldn't reach anyone. News reports showed doctors and nurses carrying newborns on oxygen to waiting ambulances for transfer to other locations. Three days after my appointment, I came to a decision. I would try to get myself in for a CVS *anywhere*, but if I failed by the end of the day, I would abort. I could not risk giving birth to a baby who carried this gene.

I remembered asking my sister how she would have felt during her own unplanned pregnancy if she ended up standing over the bed

of a dying child who was filling with chyle and slowly starving—
knowing what we know. The numbers I called referred me to other
numbers. I was offered appointments weeks away. I cried during one
call. I begged during another.

By day's end, I still didn't have an appointment for a CVS. So I
made an appointment for something else.

THIRTY

So, if I have the TIPS, can I have a baby?" I asked Dr. Sigal, after recounting the harrowing month we had endured since my angiogram. After the storm, Dr. Sigal had moved his office to another part of Manhattan.

"Yes. Absolutely," he responded.

He went back over the risks of the TIPS. I might experience encephalopathy. If I did, I could expect to feel irrationally angry and lose my sense of time. Generally, I could expect to occasionally seem out of it. The doctors assured me that the procedure could be reversed. I loved not having to commit to things.

Aaron had a lot more questions. He understood that this procedure could unblock my blocked portal vein, but he didn't like the potential negative side effects. I was feeling stubborn. I knew I shouldn't have a baby. Whether or not I *wanted* to have a baby had become increasingly beside the point—I wanted to *be able to have a baby if I wanted to*. I also wanted to prevent my condition from getting worse. Valerie had undergone a successful TIPS in 1991 when she was about my age.

"But what about her stroke?" Aaron countered.

What about it? The two probably weren't related. Or maybe the stroke happened because they removed her spleen. Theoretically, they'd never have to remove my spleen if they opened my portal vein. But it was all theory. The upside to feeling healthy while you are really, really sick is that you can take your time making decisions. That day, we decided not to decide.

A year earlier, on March 29, 2011, on her Facebook page, my cousin Denise had written something alarming:

"This has been an extremely difficult month for me," said the oversharing status update. "I need to see a liver transplant specialist." Denise, Valerie's niece, was a year younger than me. I commented. "Denise, I think you need to come with me to Boston." When she called me, she confirmed that her doctor had worried she might need a liver transplant because of a blocked portal vein.

Three blocked portal veins in the family. I decided that we needed a conference. Everyone living with our gene needed to come together at the Seidman lab to undergo testing. We needed to better understand what we were dealing with. I'd finally organized a group meeting with Dr. Kricket. Everyone came except my eighty-nine-year-old grandmother and Hilary, whose fear had gotten the best of her.

Dr. Kricket administered the tests that were required for her study, including MRIs for those who hadn't had them. It turned out three portal veins had "failed," becoming blocked. One person had minor blockage. Another had nothing. Better news came when Valerie and Suzanne learned that none of their sons had inherited our gene. Denise brought her two kids, Chelsea and Zackary, who had been born before we understood our genetic lottery. They were tested for the gene. Neither had it.

In that moment, I understood the importance of the gene not passing to Chelsea and Zack. This meant that to this point, no one in the generation after mine had the gene. It felt miraculous and empowering, in the midst of such a dire situation.

After Hilary finally got up the nerve to have an MRI done the following month in Columbus, her doctor entered the room with his hands shaking. He led with a proclamation: "Hilary, I'm afraid you are a ticking time bomb." He wasn't sure how much time she had, but after he read the alarming MRI results, there was no reason to think she had long. Moreover, she probably needed a liver transplant, posthaste. (Can you guess why? I'll give you a hint. It rhymes with "shlocked shortal shein.")

Hilary called me in tears.

What my sister's doctor told her was pretty much the same thing that Denise's doctor had told her, and even what Dr. Kricket had told me, with the gentler suggestion that I simply not throw up. I think Dr. Kricket was better prepared for what she saw when she looked at my insides swollen with collateral pathways of blood, thick snakes whirling through my stomach and small intestine. She was more ready, even if she wasn't exactly ready. Hilary's doctor, on the other hand, just got scared.

This was the moment I decided to have the TIPS. As 2011 drew to a close, I decided I wanted to have the procedure as soon as possible. Unlike Denise and Hilary, I didn't have children. I was the right guinea pig. I had Dr. Kricket's full support and Dr. Charles was ready to perform it. Aaron took several days off work after I scheduled the surgery.

Dr. Charles is a careful and deliberate surgical radiologist. When I woke up in recovery, my husband sat gently stroking my hair. A few minutes later, Dr. Charles walked into the room, smiling. The procedure had gone perfectly. As soon as he'd opened the vein, blood flowed with gusto. Aaron and I smiled at each other and sighed.

I spent the night in my own room in the NYU Langone Medical Center ICU with a stunning view of the East River. Because TIPS procedures are typically performed on the desperately ill, insurance required what was for me decidedly excessive overnight care post-procedure. ICU patients are usually asleep. Or, in other cases, like the poor woman somewhere down the hall, they are crying out in agony.

Me . . . well, I was hungry. Famished. I hadn't eaten anything since the day before. I buzzed for my nurse. He scoured the halls and finally procured the saddest-looking, yet most wonderful green Jell-O cup ever to be swallowed in one gulp.

———————

The next month I flew home to Columbus, where I accompanied Hilary for her first appointment with a specialist at the Cleveland Clinic. I wanted to be the calming presence in the room when the next announcement of imminent death was intoned. I was beginning to feel like I had become the family expert on our gene, a 360-degree turnaround since my dark days in San Francisco. I knew the options. I knew that we didn't have what our father had. I knew we also didn't have what our grandmother had. My cousin Valerie, with whom I had grown close over the last two years, was alive and thriving. We had matching flowing portal-vein shunts to prove it.

Hilary's doctor agreed that the next step was a TIPS procedure of her very own. We both felt hopeful about the future.

Later that night, I started wildly yelling at my mother. Everyone yells at their mothers, but I cried, screamed, and announced that my relationship with her was over forever. I stormed out and spent the night at Hilary's. By the next morning, I couldn't remember what the fight was about, only that we were no longer speaking. I spent the last night of my visit at Hilary's house.

As I sat waiting for my flight back to New York, I heard the announcement for preboarding. I decided to write a quick e-mail before packing up my stuff. About a minute later, I looked around. The gate was empty. My plane was gone.

Confused and panicked, I screamed at the airline employee. Why hadn't she called our flight? How could I have missed it? I was so tired and angry from the fight with my mother that I felt foggy, like I had to actively try to hold on to my thoughts or risk forgetting them. At one point, I even thought: *I have to remember I am thinking this* . . . Once I calmed myself down, I quipped to the gate attendant that she probably saw this all the time, people missing their flights.

She didn't even look up as she replied, "Not when they're sitting at the gate."

Sitting at the gate, I stared at the gangway until the next flight to New York boarded. By the following day, my anger had abated. When I recounted the trip to Aaron, he got stuck on one detail in particular.

"You can't remember what the fight with your mom was about?"

"No," I replied. "But you know my mom. She's so crazy."

"Hang on," Aaron clarified. "You mean the fight that made you think you might need to end your relationship with your mother had no starting point? You don't know why you got so mad that you left your mom's house and didn't say good-bye?"

"No," I replied again. What was he trying to say?

"And then you missed your flight, but you don't know why?"

I laughed because the idea of it was so funny.

"You don't think it's encephalopathy, do you?" he asked carefully.

"No," I answered decisively.

Quick to anger. Loss of a sense of time. Loss of my edge. Suddenly it all came into stark relief. Encephalopathy, a swollen brain caused by a failure of my liver to sufficiently clean out toxins from my blood. It had made me have some kind of meltdown.

Luckily, I had an appointment with Dr. Sigal the following week. I was feeling better. I planned on asking him: Does encephalopathy come and go? Because right then I felt like it was gone. Unfortunately, when the appointment date finally arrived, there wasn't time to ask. Dr. Sigal came barreling into the room where I was waiting patiently to pepper him with my questions about encephalopathy. He had results from an ultrasound I'd just had to monitor the stent in my portal vein.

"We need to admit you right away," he said, scaring the shit out of me. "Your TIPS is blocked. We have to clear it."

Dr. Charles managed to squeeze me in for the procedure first thing the next morning. Everything moved quickly. When I awoke several hours later, it was déjà vu, with Aaron stroking my hair in the very same recovery room I had been in the previous month. Dr. Charles came in. He wasn't smiling.

"I couldn't get it to open," he said, looking tired.

"That's okay," I said, trying to comfort him. He looked so disappointed. "Did you just reverse it?"

"What?" he asked.

"Did you just take it out? You know, reverse it?"

"I can't just take it out . . ."

"No, you can! We asked . . ." Aaron and I had both asked a number of times in a number of different ways. A TIPS can be reversed; they'd all told us.

But it turns out "reverse" didn't mean the same thing as "remove." When doctors talk about *reversing* a TIPS, they mean they block it. The little man-made piece of mesh and plastic never comes back out. It simply clots like a boulder in the middle of my liver, creating more and more varices throughout my digestive tract, waiting to burst.

Nothing was different. Actually, everything felt worse.

The next time I saw Dr. Sigal, Aaron was with me. We asked

again about whether or not I could have a baby. His answer was what I expected: carrying a baby could be exceedingly dangerous.

Aaron's reply then was short, concise, and perfect.

"Well," he said wonderfully, "it's a good thing we love dogs."

Next, we asked Dr. Sigal what his plan of action was going to be in my case. Gastric varices and an enlarged spleen still loomed ominously. He looked at us with a short, unhappy smile and shrugged.

"We wait until you bleed," he answered. That's what he said.

We wait until you bleed.

THIRTY-ONE

I bought a med alert necklace that hangs in my bathroom and only rarely around my neck unless I think it matches my outfit. It read: PORTAL VEIN BLOCK W/O CIRRHOSIS CHECK VARICES FOR RUPTURE CALL AARON, followed by his number. I don't know if Aaron would know how to explain what the words on the necklace meant if someone had to call. I don't know if anyone would understand it if they read the words. Frankly, I don't fully understand what will happen the day that necklace becomes relevant.

Dr. Sigal told me he had an idea to cut off the blood supply to my spleen in the event of a bleed. I kept planning to make an appointment with a doctor at a hospital that uses glue to stop internal bleeds. I spoke to a specialist who insisted he could still break through my blocked portal vein and ensure that it stays unblocked—although it was unclear which was bigger, his understanding of my situation or his ego. If I tried another TIPS, encephalopathy, the toxin-based brain swelling that made me miss my flight and fight with my mother, could come back. I could lose my edge. I could lose myself. Every one of my possible paths was a gamble. So I wore my med alert necklace because with combat boots and red lipstick it was a little bit punk rock.

I was still waiting for news from Dr. Kricket and the Seidman lab. I was hopeful that they would come up with an answer, that eventually they would solve our puzzle. What had caused my father's massive ascites? What had caused Valerie's stroke? My blocked portal vein?

———

Dr. Kricket sent me flowers after my TIPS procedure. When the procedure failed, she called. I told her that before the stent, when I pictured my failed portal vein, I tried to imagine blood flowing through it. I thought it might help. But now, when I pictured a boulder in the middle of my portal vein, I was scared to imagine blood flowing to it. I was afraid it would make everything worse.

"It's more like a pebble," she assured me, "placed in the middle of trickling water down a hill. The water doesn't stop because of the pebble. It goes around it."

It was a gentler picture. Dr. Kricket was a master at that. She told me that over time, she hoped, my body would reach a point of stasis, with the varices functioning efficiently and posing little danger. Hope, however, isn't the same thing as truth.

It wasn't just a sympathy call. Dr. Kricket had called to tell me about an idea she had. Actually, she said, Meredith Moore had suggested it long ago when she first mapped our gene. Since our gene was on the asthma gene, there was an active ingredient in some allergy and asthma pills on the market that she'd been wondering about. Could this ingredient trick my body out of following the orders of our gene? We wouldn't know for a while, but she wanted me to try it.

I was game. Fascinated, actually. While I was convalescing after my surgery, I'd come across a movie on TV called *Lorenzo's Oil*. I had heard of it, and possibly even seen it when it came out in 1992.

Lorenzo's Oil is a true story that sounds right out of Hollywood; George Miller, best known for directing all of the violent *Mad Max* movies (and incidentally the cartoon *Happy Feet*) wrote and directed it. It's about a four-year-old boy named Lorenzo Odone who has been diagnosed with a genetic condition passed on the X chromosome called "adrenoleukodystrophy," or ALD. ALD is a condition of the nervous system. Seemingly healthy children between the ages of two and fifteen experience sudden symptoms, usually marked by subtle behavioral changes.

Women can live unaffected with a gene for ALD on one X chromosome because they have a second X chromosome for their bodies to refer back to. Thanks to the rarity of the condition, chances are most girls will never come down with the devastating symptoms. To actually become sick with ALD, a woman would have to inherit both X chromosomes from both parents carrying a gene mutation for ALD. Ninety percent of all the people who begin to show the debilitating symptoms of ALD are little boys. One of the many horrors of ALD is that these children slowly become prisoners of bodies they can't control. Myelin, the sheath that protects the nerve endings throughout the body, begins to wear away.

In the late 1970s, when Lorenzo Odone was diagnosed, there was no known cure, or even treatment, for ALD. Unlike children with cystic fibrosis, whose parents can at least perform daily chest percussions for their comfort, people like Lorenzo's parents, Augusto and Michaela Odone, had no choice but to watch the body of their son grow increasingly devastated.

The Odones were shocked to learn that no help was available for their son. Instead, they were encouraged to join a support group for parents of dying children. Researchers by profession, the Odones undertook an extensive study of the mechanism causing Lorenzo's myelin depletion at the protein level. Over the course of a few years, they learned that the enzyme missing in ALD boys has a very similar chemical makeup to the chain of amino acids in rapeseed oil.

Although whatever myelin had already worn away could never be replaced, the Odones' discovery was a breakthrough.

The Odones developed a method to keep any more myelin from wearing away; although this would never get their son back on his feet, it would in fact save his life. Their treatment tricked their son's body out of continuing to do the bad things it was hardwired to do. They created a chemical compound that offered little boys diagnosed with ALD the chance to live into adulthood. They called it "Lorenzo's oil."

The real Lorenzo Odone never got out of his hospital bed (ALD's symptoms are irreversible once they begin) but he lived twenty years longer than he otherwise would have. He died at twenty-eight, reports say due to choking—not because of his ALD.

Dr. Kricket's plan for me to take a daily asthma pill reminded me of Lorenzo's oil. I filled my prescription immediately.

———————

Aaron had the Friday off the day after New Year's Day, 2014. We decided to visit the Frick collection, then walk down Fifth Avenue to look at the Christmas windows. Lord & Taylor's highlighted "the arts," with beautiful images like the one called "film," where a mannequin resembling Liza Minnelli twisted animatronically in front of a vintage movie camera. Several windows down, the scene in "music" cleverly displayed dozens of shiny trumpet horns that opened like metallic flowers. The Fifth Avenue jewelry stores from Tiffany to Bulgari glittered with millions of brightly shining lights.

After spending nine dollars on a bag of dried-out chestnuts and a bottle of water, we decided to head back to Brooklyn. As we stood on the F-train platform near Rockefeller Center, Aaron suddenly turned to me and stared for a minute. He smiled and said out of the blue, "You know, I love you very much. I love our life and I'm happy."

He could be effusive at times, but at this moment my stomach filled with sixth-grade butterflies and I began to blush. Then, suddenly, I felt sick to my stomach. Rivaling his romantic proclamation with one of my own, I answered dryly, "I just really hope I don't die."

"Jesus, well, yeah," he agreed, modulating his voice to the screech of metal and breaks as the train pulled up in front of us. "I've never had a disruption in my life like that, like you have."

We boarded the crowded train and stood with our faces close together. I was suddenly desperate to make an important point that had occurred to me. I squeezed his hand and looked into his eyes.

"Promise me," I began. "You will not fall apart." I stared at him hard. "I need you to promise me you will go on and live a full life if I die."

He matched the seriousness of my gaze for a moment and then his face settled into a faraway stare. A smile played on his lips as his expression turned wistful. Then, as if he suddenly remembered I was standing there, he quickly cleared his throat and squeezed my hand. "Right. Of course, baby. Somehow . . . after a respectable mourning period, of course . . ." He cleared his throat again. "I don't know how, but I'll try and go on."

I laughed all the way to Forty-Second Street.

THIRTY-TWO

Aaron and I scheduled a trip to Boston in March of 2014.
When we arrived at the Brigham, Barbara, the lovely
nurse who worked with the Seidmans, sat with us in a room. Soon
Dr. Kricket and a new postdoc fellow, who had taken over for Meredith Moore, joined us. Aaron and I sat at a round table, flanked by a
laptop computer and a big-screen monitor. The room was small and
felt crowded, but the atmosphere was electric as Dr. Kricket began
her presentation.

My father had worked so hard to understand his condition
twenty years ago. He had tried out every possibility. Now he was
gone, but every one of his questions still lingered, unanswered. In
fact, seventy-five whole years had come and gone since my great-grandmother left her husband and her five young children to live on
a screened-in porch so that she could figure out how to take in a full
breath of air. Lymphedema had filled the limbs and bodies of five
generations of my family, and had already killed five.

Dr. Kricket began to speak.

"We're going back to ancient history, my friend," she began, addressing me as she launched into her presentation full steam ahead. First, she reexplained the job of the lymphatic system.

"The lymphatics," she explained, "are a salvage pathway. They pick up all sorts of material—debris, fluids, electrolytes—that seep out of vessels during their normal processes." She went on to say that initially, she and her team had looked into several very rare conditions that cause lymphedema. None of them fit, and she knew this largely because of the fluid that was present in my stricken family members—it wasn't yellow and thin like the fluid in a blister; it was thick and sticky, lymphocytic exudate.

The next step was to look into the mechanism that channels the fluid, the lymphatic system itself. It was easy to rule out a broken lymphatic system, largely because of the condition's adult onset. If the problem originated in the vessels of the lymphatic system, those of us carrying the family gene probably would have had abnormal swelling our whole lives.

"If you think about it," she told us, "if you guys didn't have working lymphatics, why would you start having this problem when you were forty? You would have had this problem when you were a kid. Kids get banged up all the time. You'd have swelled with every bruise. And you didn't. So we didn't really expect a broken lymphatic system."

When my great-aunt Norma died, Dr. Kricket conducted her autopsy and ruled out any abnormalities of the thoracic duct, the entry vessel between the lymphatic and the vascular systems. "Even the anatomist at Harvard Medical School, who analyzes autopsies for a living, came and found no abnormality there," she said.

At that time, she and her staff had begun talking to everyone in the family, and the first thing they discovered was that no one under the age of twenty had had any excess fluid anywhere. What most of us did have were heart murmurs, indicating a narrowing in a vessel of our hearts. "A little narrowing," she said, "above the pulmonic valve."

At this point, Dr. Kricket put up a family tree—called a "pedigree" when used for medical or scientific studies—on the screen. It included every family member with the gene, and the symptoms they had exhibited.

"So the reason I've put you through this torture," she said with a wink, "is to tell you, we have found a very, very, very rare variant and it has never been described in anybody else."

The enormity of that statement momentarily made me dizzy. Never. In anybody else . . . except *my* great-great-grandmother, *my* great-grandmother, *my* grandmother, *my* great-uncle, *my* great-aunt, *my* father, *my* uncle, *my* sister, *my* five cousins, and myself.

Twenty years ago, in one of the medical journals my father kept, he had written that there were "no additional recommendations or advice" because "no one knows what we are dealing with." It had been a very long line of nonanswers. From my great-grandmother Mae's misdiagnosis of pleurisy to my great-uncle Nathan's ten months at the NIH that ended with a chart note that read "the basic disorder . . . is, after all these many months of hospitalization, still completely unknown."

Dr. Kricket was about to change all that. Once again, she was about to change everything.

One little note in one of the many lists of questions from my father's personal medical journals simply reads, "It can't be just the liver, can it?" He wrote exactly that.

It can't be just the liver. Can it?

It was almost like he had to make the definitive statement: *It can't be just the liver.* But then he'd thought about it for a second and wondered, "Can it?"

Dr. Kricket began the rundown of what her lab's extensive, twenty-year study had revealed. It was this: *our condition was vascular.* And that "vasculopathy," as it's called, seemed to be originating in the liver. So yes, to answer my father's question: it *can* be just the liver.

Everyone's liver makes something called cysteinyl leukotrienes (CLs). Those in my family do it too. We all have these things in our livers called cysteinyl leukotriene receptors. The CLs are the keys and the CL receptors are the locks. When our keys get to our locks, they are supposed to unlock them. If they listen to our mutant gene, however, they will do something abnormal.

"We have a very small idea about what and why that is, but I don't want to drive you crazy," Dr. Kricket added. "What we do know is that that system regulates *pressure*."

There are blood pressures in our liver that pull blood from our intestines and our spleens in order to purify it. That's what our livers do—they clean our bodies of toxins. But in my family, a mutant gene causes liver pressures to become confused and the liver to no longer be able to properly siphon the toxins out of our blood.

"We think that fundamentally there is an abnormal signaling pathway that changes the pressure," she continued. "As a compensation, there is an expansion of vessels to get around the liver to get the blood so it circulates normally. The spleen gets larger, the lymphatics get compressed, and you end up with lymphedema."

It's this little glitch that causes this complication throughout our bodies, and it is very likely impacting other parts of our vascular system as well. This includes our heart murmurs; my dad's sticky heart valve; my great-uncle's drooping eyelid, caused by a collapsed vein in the back of his eye; and even my cousin Valerie's stroke. When it does its worst, it tackles our lymphatics, knocking out our body's ability to absorb nutrients, causing us to slowly starve.

And that is our breathtaking little gene.

After Dr. Kricket's presentation, Barbara, the nurse, took my hand. I had known her for half of my life. She is one of Dr. Kricket's right hands, juggling several of the Seidman "balls" at any given time, and always with smiling dimples and ruddy red cheeks. Her red hair has grown blonder the longer I've known her, while she somehow manages never to age.

"I want to show you something," she said.

I had only been to the lab once before. It was a crowded, active space, with papers cluttering desktops covered with microscopes and old computers. Shelves housed piles of equipment I'd never seen before and some that took me back to my high school biology class.

We walked into another room and then to a darkened room several turns from where we had begun.

"I thought you might like to see this," Barbara said, smiling. She pointed to a small sturdy metal cylinder about four feet high. It reminded me of the home kiln one of my mother's best friends, a clay sculptor, kept in her basement.

"It's a freezer," Barbara said, the apples in her cheeks getting redder. "It's where we keep some of our immortalized cells . . ."

These weren't just any immortalized cells. These belonged to my family.

"Your dad's alive in there," she said.

I felt the hair on my arms stand up. My father had been dead for eighteen years. I smiled back at her as she squeezed my hand, tears filling our eyes.

Stranger still was my next thought: *I'm alive in there too.*

———

The meeting at the Seidman lab had energized me. All we had to do now was go to a pressure doctor and figure out how to maintain the pressures that were off in our veins. Or, as I so readily have taken to putting it at dinner parties, "We just have to figure out how to keep our veins pumping correctly."

At one such dinner party, however, someone quickly burst my bubble.

"But veins are passive," said a guest who knew what he was talking about.

I mumbled something about how delicious the salad was.

That person was right. Veins are passive. They don't really *do* a whole lot. They are channels, receivers. It's a little bit like saying that the congestion in the Holland Tunnel could be fixed if we could just get the Holland Tunnel to *push* the cars through, gulp them down into New Jersey. Veins don't pump.

What's more, there aren't really a great number of vascular specialists out there, largely because . . . and I hate to be redundant, but *veins are passive.* Most vascular specialists specialize, much like an engineer does with a tunnel, in helping to make sure the veins are routing things properly. The biggest job of most vascular specialists isn't to worry as much about the stuff being transported—the blood, the lymph, the waste—but to worry about *where* it's being transported.

This brings us back to cardiologists. These are the people who are in charge of the thing that pumps—the heart. Remember, according to Dr. Kricket, there was something off in the pressures of our veins. Cardiologists, like Dr. Kricket, are on very intimate terms with things like pressures because they are focused on the thing that is exerting the force that is making the blood run through our veins—and arteries and lymphatics. Cardiologists know tons about the heart. As opposed to the vascular and the lymphatic systems, they know loads about the arteries—the first and largest channel into which the heart pumps. But—and this is not to suggest that cardiologists aren't brilliant—as the arteries meet veins and blood moves onward, their area of expertise typically diminishes.

Surgical radiologists spend a lot of time with veins. They love that veins are passive. They especially love that veins are passive and the pressure of the blood flow in them is much lower than arteries. It means they can explore a patient's body with a camera, or guide a catheter or other device from outside of the body to a place requiring a corrective procedure without ever having to cut into a person's flesh. "Surgical" radiologist is actually a misleading term, because surgical radiologists actually make what used to be a surgical procedure *less surgical.* Today, my father's zipper scar from neck to navel would never have been cut. Instead, a small prick at the jugular vein in his neck, or in his groin, would allow a radiologist to drive a balloon through the veins to his heart and correct the narrowing by blowing that balloon up when it got there.

Again, radiologists aren't vein specialists per se.

My friend Molly, the nurse who introduced me to the idea that a burst varice in my esophagus could lead to a bleed out of my entire blood supply through my mouth, is a PICC-certified nurse. "PICC" stands for "peripherally inserted central catheter," and is pronounced *pick.* This certification actually allows a nurse or technician to access any part of a patient's vasculature. Maybe a PICC-certified nurse is as close as we will ever get to a vein specialist. The key thing to note here is that these nurse specialists pretty much always access *veins.* Almost all of their procedures require an extensive and thorough understanding of veins and vein pressures. Like radiologists, they choose veins because veins are the best highways to navigate through a body. They have a relatively strong structure like an artery, but they don't have blood flow like a barreling freight train—more like a slow and steady hand-crank locomotive.

Here's how the circulatory system basically goes. The heart pumps a giant squirt of clean, freshly oxygenated blood right into an artery and it moves through the body, delivering oxygen and nutrients while collecting toxins to flush out. As Molly explains, when she's threading a wire to get to the edge of the heart, she knows she is in an artery when she loses control. She often has to pull the wire back out and start again. When you consider that the whole of the circulatory system is sixty thousand miles long—that's two and a half times the circumference of the earth—you realize just how much work the heart has to do. With every new pump, it has to start a new cycle and push forward the rest of the blood supply already en route. However, it's safe to say that pressures close to the heart are much higher than those in, say, your toes.

Arteries carry all the good stuff from your heart and through the body. Your veins, meanwhile, are in charge of the return trip to the heart. They rely on whatever pressures are left after pushing blood through a very long and arduous journey through the arterial system.

That isn't all there is to it. Arteries are not *passive*. They actually provide resistance against what is hopefully a strongly beating heart. They aren't just Holland Tunnels receiving a bunch of cars. They are helping to control the pace at which those cars move, pushing back—like the Highway Patrol—to keep the speedsters from driving at maximum speeds. The thing is, veins aren't entirely passive either! Yes. I know what I said. Imagine if our veins just sat there, receiving blood and lymph and waste, and didn't *do* anything. The vein would simply expand and expand, when it actually needs to *channel* blood. Even the veins, albeit at a diminished rate compared to the arteries, help control how much and how fast the blood moves through them. In fact, they help to control the pressures. I suppose they're surprisingly active.

So, could it be that the blood pressures that are controlled by the veins—minimal though they may be—are what was causing the blockages and lymphatic overspill in my family members? The eyelid drooping and the strokes? The murmurs and the enlarged spleens? Was I right at those dinner parties? The veins weren't pumping correctly?

Dr. Kricket doesn't think so, and she said as much during our Boston meeting, although it took some time for her words to settle in. Dr. Kricket said it: cysteinyl leukotriene receptors manage pressures in our *livers*. It turns out that Dr. Kricket now believes that something else besides the heart pumps blood: our livers. While studying our gene, she has been able to explore the fact that livers also help pump blood. This revelation is a pretty big deal.

"The liver," Dr. Kricket said that day, "is making molecules that regulate blood flow into the liver." The protein that serves as a sensor for those regulatory molecules mutated in us in our bodies, on a gene in our X-chromosome where that protein doesn't exist.

The liver plays a very important role in our bodies. As I said before, it purifies and removes unwanted materials from the blood. Food goes into our mouths, and after hitting the stomach, it gets absorbed into our bloodstream. Then those nutrients are taken to the intestinal tract and are processed into waste. A vein called the "*portal*" vein brings the blood into the liver so it can clean and detoxify whatever it has to clean and detoxify. This is an important function that keeps our bodies healthy, and it turns out that the liver helps regulate the flow of this blood: a surprising revelation. My family's livers, because of our genetic mutation, were not correctly regulating the blood flow into them through the portal vein. In my family members' bodies, the portal vein changes, withers, as a result of "weird" pressures causing blockages. Eventually, everything backs up.

The Seidman lab was trying to figure out whether or not the pressures in the portal vein could be increased. In fact, a number of bioengineers are studying this question. At least, that's what Dr. Kricket has told me.

A cure for us remains hypothetical. It is far from guaranteed.

But when you're still feeling healthy, "hypothetical" and "far from guaranteed" don't add up to "out of the question," which makes a cure almost seem possible—even likely. When you are no longer healthy, "far from guaranteed" becomes about the same as "out of the question."

My father knew this well.

"I am sure that if we get a diagnosis that's complete," reads the final sentence of an entry in his medical journal, "the treatment will not be easy."

THIRTY-THREE

It is rational to fear death, but it is not logical.
—*SOCRATES*

My stroke started on a Sunday during a road trip. It had been a full year since Dr. Kricket had figured out the minutiae of my family's genetic condition. I was staring at the clock on the far wall as I sat at the desk of the emergency room admissions clerk at a hospital in Hartford. It was taking too long. My head was filling with blood. I could feel it. *I'm bleeding.*

"I'm bleeding," I told the clerk outside the ER.

Aaron was next to me.

We'd left our car outside the ER door. We didn't care that the signs said NO PARKING. A uniformed police officer had come in and was now talking to Aaron. Maybe about the car. Aaron got up and followed the police officer outside, perhaps to move the car.

Aaron is going to be alone, I thought. *Does he know about the yams? Did I tell him that the yams are really important to me?* I really wanted

our dogs to eat a scoop of boiled yams with their kibble. I couldn't
remember if I'd ever said it.

*Oh, shit. What if, when he gets back, I'm too out of it to explain it to
him?* The blood was filling my brain. Maybe my body cavity. How
would I explain the yams? I could barely speak. It's all so white—
everything in my line of vision was covered in a white foamy haze.
What if all I could say was "Yams," and then I died, and then Aaron
had to be like, *What the hell did she mean?* What if "Yams" became my
"Rosebud"? Would that just be the lamest thing? Did I really care if
the dogs ate yams? How much did I care? I tried to quickly figure it
out before I was too far gone.

My thoughts were, I'm sorry to admit, mostly about yams. I told
the admissions clerk about the blockage in my portal vein. I told her
that my body was filled with varices. "Gastric varices," I told her. "I'm
having a bleed."

She looked at me kindly. She was too young to actually have a job
at a hospital. Did she even know what a gastric varice was? A delicate pathway that could burst and cause internal bleeding? Could
cause a stroke? Could pour out my entire blood supply through my
mouth?

I was shaking. I called Dr. Sigal, my doctor at NYU.

"I'm in the ER," I told him. My voice was so far away. He told me
to have the doctor call him as soon as possible.

Aaron was back with me now. He'd moved the car. I was mulling
over the details of my life. What else did I wish I had taken care of
before I died? Before my brain fell apart?

In fact, in the midst of everything, I realized one important thing:
I'm not very important. Maybe that wasn't what I meant. I knew, like
Jimmy Stewart had known, that Bedford Falls was better off because
of him, and all that—but I was so mentally prepared for the moment
that I would no longer be, that my thoughts, yams included, were
fairly mundane.

I worried for my husband. I worried that he would be consumed by rage. I worried that he would embrace his loneliness and isolate himself from the rest of the world. I worried that he would miss me as much as I would miss him if anything were to take him from my life. When I think about it, I am overcome, because I hate imagining that pain.

I worry that my dogs will never understand why I left them. I worry that my nieces and nephews will read something I've written and misunderstand me and never get to ask me what I meant. I worry about these things. But there is no way to prepare for them.

The ER doctor came into the room and took everything I said very seriously. She nodded and wrote things down. She touched my wrist, feeling for my pulse. I was hooked up to machines and blood was drawn. I could still feel the searing-cold burn in my limbs, but the fog in my head was lifting. An intern asked me to raise my arms and touch my nose with a finger on each hand at the same time.

"If you were having a stroke," she explained, "you'd typically have different movement on each side. Or one side might be harder to control than the other."

Based on her logic, it seemed I was not having a stroke. Both of my fingers met my nose without confusion. But the searing. The haze. No, something was still definitely off. It was probably an internal bleed. I'd been coughing from a summer cold that wouldn't go away. Each cough was violent and I couldn't control them. I'd taken cough syrup but the coughing didn't stop. I knew that coughing could cause a bleed. That involuntary spasm, the throaty scrape. It wasn't good.

In the gut-wrenching book *Paula,* Isabel Allende writes about the tragic and sudden death of her twenty-nine-year-old daughter. One line in particular resonated with me: "Life is nothing but noise between two infinite silences."

I thought that no one had ever put the truth of the matter so succinctly and beautifully. The thing is, I have always been living-centric. I think we are disposed to focusing on the times before and after living—because the living part is so short, so noisy. I understand the need to focus on those who have passed and those who are yet to come. Everyone we love, have loved, or will love has and will spend most of the time *not living*.

I understand that having a belief in a place for the dead and pre-born is part of the fabric of almost all religions. I hope it's true, at least the stuff about the good places for the dead and preborn. Still, I am an optimist and am disposed to loving my here and now, which makes the living, and their ever-changing needs, the most appealing part of life to me. I care far less about what came before and what will come after.

As I lay in that hospital ER dying, it became very clear to me that we must each make our peace with the mysteries we will die never understanding. We must be comfortable with the fact that we won't know what the wrinkles will look like on our children's faces when they are old, or how robots will take over the world—or become really good servants—or the conclusion to *Game of Thrones*. We must also make our peace with how each of us will be remembered, or forgotten.

The thing was, I had just read a really good horoscope for August, and now I was going to die. And if I didn't die, I was probably going to be severely impaired. This moment: *this*. It was the beginning of everything finally changing.

Although my affairs were more or less in order, I was terrified.

The ER doctor came back in. She'd spoken to Dr. Sigal.

"We've done all the testing he asked us to do," she explained. "Everything looks good. But stay here for another hour. We'll check one more time."

The ER nurse who had been hovering nearby since I'd arrived explained that my blood draws had suggested I was not internally bleeding. If I were bleeding internally, my blood levels would be low, but they were not. So she asked me about my weekend.

Aaron and I had driven to Boston for my cousin Becky's birthday, I told her. It was just an overnight. Becky's young, vibrant husband, Peter, was dying of cancer. He was swollen from his steroid cocktails, but still beautiful. Becky was trying so hard to act the hostess, helping her frail husband down the stairs. She was full of desperate laughter and kind attempts to make everyone feel at home, to make it all feel normal. Her sister and mother had prepared all the food. But Peter was still dying, and he was too young to die. Becky just wanted one night of normalcy, and all of us were trying to give it to her as best we could.

The next day, as Aaron and I drove through Hartford on our way home to Brooklyn, we ate at an Olive Garden. We lamented how it is the one restaurant that always sounds better than it actually tastes. Afterward, we decided to stop at the Mark Twain house for a tour. I was coughing, thanks to my summer cold. Then my head filled with feathers.

"I think I have to go to the hospital," I told Aaron, quickly typing *Emergency Room* into my iPhone map app. Siri told us to return to the route.

"Drive slowly," I told Aaron. "Maybe I'm all right."

"Okay," he replied. He was quiet, looking over at me occasionally, steady and calm.

The ER at the Hartford Hospital is a lovely place to fall apart. When the ER nurse asked me if I'd ever had a panic attack and I said no, she replied, "Oh, honey. If I had all the problems you described today, I would have had twenty by now. You were right to come."

After my second round of tests, I was discharged. My husband helped me take off the hospital gown and put on my clothes.

I was alive.

I didn't care that it was a panic attack. That I had bothered my doctor on a Sunday. That I had missed Mark Twain. I was alive.

The normalcy of the next few days overwhelmed me with their beauty.

"I love you so much," I told Aaron ninety times an hour. I hugged my dogs until they pulled away and hid under the bed.

I made a lot of yams.

I suppose I was bound to crack sooner or later. *Next time*, I vowed, *I will understand my panic for what it is.* Unless next time is the moment when everything changes.

THIRTY-FOUR

The world of genomic medicine is a promising place.

In the first half of the year 1999, gene therapy was all the rage. Articles touting the wondrous possibilities abounded in the media. Gene therapy was the panacea for everything from hemophilia to cardiovascular disease. It promised to change medicine. It promised to change the world. Using viruses to reconfigure genetic code, scientists theorized that they might actually be able to reprogram genes. It was an idea so fantastic that money poured in from everywhere in the hopes that gene therapy might someday cure everything from cancer to old age *by changing our genes.*

Then the unthinkable happened: on September 13, 1999, Jesse Gelsinger, a relatively healthy eighteen-year-old, arrived at a hospital for an experimental injection. Jesse had a mild form of a liver condition that kept sufferers from metabolizing ammonia, called "ornithine transcarbamylase deficiency." He was participating in a gene-therapy trial. In infants, OTC deficiency nearly always proves fatal. The first gene-therapy trial in a human took place in 1990, and early testing was looking good. Jesse, who had been required to follow a special diet and take a steady dose of thirty-two pills a day for most of his life, was happy about the potential to rid himself of the illness. Moreover, he was happy to be part of something that could change other lives for the better as well.

By the time this treatment became available to Jesse, its safety was a foregone conclusion. Animal models had already proven that any risks were limited. The administering doctors were confident that Jesse would not only survive the treatment, but he would benefit from it.

Jesse received the injection. Four days later, he died. His official cause of death was listed as lung failure. However, the treatment Jesse had willingly subjected himself to ("for the babies," the *New York Times* reported he'd told a friend) was something out of science fiction. Corrective genes were infused into a common-cold virus functioning as a "vector." Vectors are the transportation devices that scientists use to drive DNA into cells and reprogram them. Researchers typically use them in order to change good information to bad—for example, to mutate the genes of mice in order to study human genetic illnesses. In Jesse's case, they hoped these vectors would change his DNA from bad to good.

Jesse's death had a ripple effect. Medical studies involving gene therapy were suspended and in some cases stopped completely. For the next ten years, gene therapy and all the wonders it had promised screeched to a grinding halt, or so it might have seemed from the outside. In private labs, far from media scrutiny, gene therapies continued to be tested. They progressed at a much slower pace than before, but they were still being researched, and slowly, steadily, they were beginning to work.

Fourteen years after Jesse Gelsinger died, three-year-old Eliza O'Neill was diagnosed with a recessive condition called Sanfilippo syndrome in 2013. After developing normally, at the age of two, Eliza began to regress. If her illness were left unchecked, she'd eventually lose the ability to speak and walk. Her parents, Glenn and

Cara O'Neill, had only one option: prepare to watch their daughter die. The O'Neills decided to change that fate, for their daughter and every other child suffering from Sanfilippo's.

Cara was a doctor and Glenn had a background in business. Upon learning that there was a rudimentary gene therapy for Sanfilippo's that had shown itself to be successful in models using mice but lacked the necessary funding to enter a human testing phase, they decided to raise the money themselves. The first thing they did was quarantine their family of four. If Eliza was going to remain eligible for the trial, her parents had to keep her from exposure to the AAV9 virus, which would exclude her from eligibility.* During the quarantine, her disease would progress, but her parents intended to do everything possible to keep their child eligible for the trial, so the family of four, including her now-nine-year-old brother, Beckham, stayed in virtual seclusion from the outside world for 726 days.

Sanfilippo's is a "lysosomal storage disease" in children that causes severe neurological degeneration and early death. A part of the cell designed to recycle specific molecules fails because a mutation robs it of one of forty-three necessary enzymes. A buildup of a substance called "heparan sulfate" causes all systems to eventually break down. Eliza was, at the present moment, as well as she would ever be. The O'Neills needed her to receive the treatment before things became worse, before she lost her ability to speak or other cognitive capacities. The clock was ticking.

After Jesse Gelsinger's death in 1999, many people were skeptical about gene therapy. In the summer of 2014, as the O'Neills barricaded themselves inside their home and prepared to fund-raise for a gene therapy that could save their daughter and others like her, something remarkable happened. Within eight months, their campaign, mostly limited to social media, had amassed 22,337 individual

---

* AAV9 has become a go-to virus for gene-therapy delivery and was the planned delivery method for Eliza's trial. Exposure would render her ineligible.

donors with donations totaling $1,040,000. That number is now in the multiple millions, and the foundation started by the O'Neills is finally funding urgent research.

In May of 2016, Eliza was the first child to undergo a gene-therapy treatment for Sanfilippo syndrome. With a onetime injection, a corrected gene was sent directly into her bloodstream. Researchers are hopeful that Eliza's outcome will mirror that of the mice models, which included restored cognition, improved mobility, and best of all, an extended life. In just a few months' time, the O'Neills became hopeful about a bright future for their daughter.

Gene therapies do one of three things: they replace a mutated gene with a healthy version of the gene; they "delete" a gene that is causing problems; or they introduce a new gene into the body, typically to fight a specific disease. In all three cases, genes are manipulated. Today, these therapies are only being tested on those patients who have no other options. Patients essentially have to be on their way out to become a part of one of these trials. What's remarkable is that the trials are showing signs of working.

Color-blindness is the basic inability to perceive color. Typically, a color-blind person can see some but not all colors. One gene for color-blindness, like my family's gene, is on the X chromosome and is recessive. If an XY male gets an X from his mother with the gene for color-blindness, he will be color-blind. An XX female must get the color-blindness gene from both parents. Color-blindness isn't deadly. It doesn't make you swell or deplete your myelin sheath or fill your lungs with fluid. There was no cure for it because it was in the thirty-seven trillion cells of those who suffer from it.

In 2009, that changed. Doctors at the University of Washington led by Dr. Jay Neitz came up with a revolutionary gene therapy. They injected a gene that produces a protein missing in color-blind rhesus monkeys directly into the monkeys' eyes. Right now. Today. Gene therapy is changing something that was once an immutable fact—it is changing genes!

Rhesus monkeys are a popular test subject for human medical genetic treatments largely because they share so much of our genome. The color-blindness gene, called "L-opsin," when missing, eliminates red and green from the perceived color palette of both people and monkeys. Dr. Neitz doesn't have all the answers yet, but five months after injecting L-opsin into his monkey subjects' eyes, these animals suddenly had the neurological capacity to distinguish red and green. He believes it has something to do with their brains reconfiguring themselves to contain the new information in the injected gene.

This therapy has blown open the door for treatments of multiple forms of congenital blindness. For example, Leber's congenital amaurosis, a form of blindness caused by a breakdown in a sufferer's ability to perceive light, has long been a contender for successful gene-therapy treatments.

An even better contender for making gene therapy viable is a process called CRISPR. The acronym stands for "*clustered regularly interspaced short palindromic repeats*." What CRISPR technology does is *edit* genes. It is like gene therapy—or *is* in fact gene therapy—except that the results of CRISPR are reliable and predictable where previous methods were a little more, shall we say, finger-crossy.

In an episode of NPR's *Radiolab* called "Antibodies, Part I," hosts Jad Abumrad and Robert Krulwich very clearly articulate the amazing and wholly terrifying technology that is CRISPR. It all began when scientists who were reading the genetic code for some bacterial strains noticed short, strange sequences that, to the best of their knowledge, shouldn't have been there. In the *Radiolab* episode, this situation is dramatized by having a single saxophone note sound repeatedly and suddenly tossing in a strange off-pitch note while the saxophone note continues. (Hence the name "clustered regularly interspaced short palindromic repeats"—CRISPR.) Biologists believed that this jarringly odd note in the middle of the series of expected notes had to have a purpose. What was it? Or, as Abumrad puts it, "Now scientists had this puzzle. If nature is preserving something at this level, you figure, well, whatever this is, it's doing something."

A biologist at the National Center for Biotech Information named Eugene Koonin suggested it was in fact doing something: it was acting as a defense system. To continue with the musical metaphor, the odd note imposing itself into the middle of a sequence of bacterial DNA was virus DNA that was being stored "*like a mug shot,*" Koonin says in the episode. You've probably heard of something like this in explanations of the phenomenon that our bodies tend not to get the same virus twice. This reasoning explains how immunizations work. Our immune system stores viral information so that it is prepared to fight that virus again later if it has to. With CRISPR technology, we can arm our own immune systems with a virtual mug shot, allowing them to recognize and target flawed genetic information with incredible specificity. "Immunotherapy" studies for cancer treatments are one area where we are already benefiting from this remarkable technology.

What happened to Jesse Gelsinger wouldn't have happened if CRISPR technology had been available. It's the most powerful gene engineering we've ever had. It's cheap and precise. Using CRISPR, we might be able to take the conversation from "How do we prevent global warming?" to "How can we alter the genes of corn so it needs less water in order to grow?" Or imagine answering this one: "How can we alter our genes so we can comfortably live underwater?"

Of course, it is also possible that the powerful tools gene therapy is discovering may be beyond our understanding. Whether we are an empathetic enough animal to use CRISPR humanely or in humanity's best interests remains to be seen.

While these treatments make me feel hopeful for myself and my family, right now they remain too costly. The Seidman lab, though, is still working on our case. To this day, they haven't published their findings, and they haven't yet given a name to our disease, but they're working on it. In the fall of 2015, Dr. Kricket told me three things. The first was that three mice had been mutated with our gene. The second was that two of those three mice had already died.

The third thing Dr. Kricket told me was that she was planning to do something amazing with my blood cells, if I were willing to sign a consent form—which I am. They are going to turn them into "*induced pluripotent stem cells.*" In 2012, Shinya Yamanaka of Japan and John B. Gurdon of England won a Nobel Prize for the discovery that adult cells can be reprogrammed into something like stem cells. The word "pluripotency" refers to a cell's ability to develop into any of several types of cells. For instance, a liver cell, once it has become a liver cell, stays a liver cell. A blood cell, once it's a blood cell, stays a blood cell. This is why the discovery of stem cells in embryos in 1981 was so exciting. They were cells that could be induced to grow into a blood cell, or a liver cell. In fact, there was a time when people thought that one day stem cells might be removed from developing fetuses and harvested for the people who grew out of those fetuses, should they ever need a liver transplant or blood transfusion. They'd actually be able to get their own blood or a new liver, whatever they needed—created outside their own body from their own stem cells.

Yamanaka and Gurdon went one better and figured out how to turn adult cells into stem cells. At a molecular level, the cell is genetically reprogrammed to return to a stem-cell state. Dr. Kricket said, "We can actually take a blood cell from you and make it go back in time!" She said it can be turned into an early heart cell, an early liver cell . . . and on and on.

Dr. Kricket hopes her team will be able to prove that my family members' livers are making molecules that impact our vascular liver pressures because of our genes. She wants to use these cells to grow vessels to make proteins and see how normal proteins stick to them. This is a different way of proving what she already knows about our gene.

In the winter of 2016, the last remaining mouse with our mutated gene successfully bred. Today our gene lives on in multiple mice in a basement enclosure of the genetics department at the Harvard University Medical School. A mouse's life span is 1.5 to 3 years. Hopefully in the next few years, we will see a mouse model develop our vascular condition, proving that our gene is where Dr. Kricket says it is, as well as all of the theories she's developed during twenty-five years of hard work.

It's more than my dad ever thought possible. It seems like his unwavering faith in medicine is being rewarded at last.

THIRTY-FIVE

Aaron and I didn't have children. It's true that medicine has come far enough that if we decide we want to, we can. We would find a way to raise the necessary funds and hire a surrogate to carry a genetically related fetus. We could utilize a "preimplantation genetic diagnosis" before selecting an embryo that does not carry my family's mutant gene. In this way, we might have a baby. But Aaron and I don't feel any pressure to have children yet. We like our respective sisters' children so well that maybe we feel like our collective gene pools might have already done their best work.

We talk about adoption or fostering. Again, we have not begun to take the necessary steps to do either. For most people, the issue of parenting is not as fraught as it is for Aaron and me. There's a broken condom or a night of drunken passion. There's enough mindless sex to turn the vague notion of having kids into a very real bawling baby.

I suspect that I would be a good mother and that Aaron would be a fantastic father. I understand that parenting deepens you and expands you. My mother regularly tells me that she feels sad for me when she considers that I may never be a mother. I am also, as I've said before, an incurable optimist, annoyingly so at times. I tend to like my life exactly as it is. Sometimes I think it would be great to add a child to my happy family. But then again, I haven't done much about adding one.

There are moments when I think I am selfish for putting Aaron in a position where he will be left alone if one of my varices bursts and I die or I become mentally deficient. Technically speaking, I am more likely to die, or grow seriously impaired quickly, because of a compromised vessel than because of a slow and debilitating lymph accumulation like my father. When I consider this, I think I would be more selfish to bring a child into a family where such a monster lurks.

Of course, these thoughts swiftly disintegrate when I remember that love is always a gamble. Though children might make us better, we also risk the agony of their loss, whether or not your portal vein is blocked, or your genes are faulty.

Each of us. Every day.

My friend Elle understands this better than pretty much anyone I know. Elle, with her dark hair, feather-white skin, and doe eyes, is charmingly beautiful. She is so shy that when she speaks her cheeks redden. If Elle's genome has anything to say about it, she is going to die young. Right now she is in her mid-thirties, and her body is poised to start breaking down at any minute. Still, she and her husband just had a baby boy.

When Elle and John met six years ago, they didn't fully know about Elle's future. There had been a smattering of cases of "Huntington's disease" (formerly known as "Huntington's chorea") in Elle's family tree. No one talked about the fact that it was genetic. At the time, there wasn't much that could be done about it. Huntington's has an adult onset. Sufferers typically don't fall ill until they are into or even beyond their childbearing years. At a molecular level, adult-onset illnesses typically involve a slow breakdown.

Dr. Kricket has discovered that the bodies of my family members have too few cysteinyl-leukotriene receptor keys in their livers to match up with their cysteinyl-leukotriene locks. As we age, her theory is that the number decreases further until eventually there are not enough keys for our locks. This is why we aren't born sick. Why we don't grow sick until we're past the age of childbearing.

The gene for Huntington's causes an abnormally long version of a protein known as the "Huntington protein" to develop. As the protein breaks into toxic fragments, it binds to neurons, affecting cell functions. This toxic buildup causes Huntington's early symptoms, but over time, the neuron dies. Soon after, so does the person. Unfortunately, with every generation, that Huntington protein gets longer, and the life span of the sufferer is shorter.

Perhaps the most famous victim of Huntington's was the folk singer Woody Guthrie. The "This Land Is Your Land" writer and singer began experiencing early symptoms of the illness in his thirties. By the time he was forty, he was correctly diagnosed. However, by then he had already been labeled both alcoholic and schizophrenic. Mental health issues are common for sufferers of Huntington's, as the disease is neurodegenerative, meaning that it leads to a loss of neuron function.

Guthrie's mother, Nora, had Huntington's. During her decline, she set two coal-oil fires, one that killed Guthrie's seven-year-old sister and another that disfigured his father. Over time, sufferers can spend years without speech, steadily losing bodily control until they die.

Woody Guthrie had eight children, including a son, Arlo. Arlo grew up to be a famous musician like his dad. His hit "City of New Orleans" was one of my parents' favorites. *"Good morning, America. How are you?"* went the chorus. *"Don't you know me? / I'm your native son. / I'm the train they call the City of New Orleans. / I'll be gone 500 miles when the day is done."*

During an intro-to-psychology class I took in 1994, the professor alluded to the Guthries. "Arlo Guthrie," she said, "can't know if he has the gene for Huntington's disease until he gets sick with Huntington's disease." She pointed out that this would factor into the younger Guthrie's reproductive choices as well as into his other life choices.

Perhaps at the time I was hearing about him, Arlo couldn't find out if he had the gene for Huntington's until he did or didn't come down with Huntington's, but I found his story fascinating. There was something so inherently cruel in the kind of lottery his fate had caused him to play. Go ahead and live your life, Arlo Guthrie. But don't get too comfortable, and stay away from coal oil and matches.

About fifteen years ago, Elle's father lost his job. Elle was away at college at the time, but she says he was having trouble keeping up at his new job—which the family mostly attributed to the fact that it was a new job. There were little things that caused contention, like forgetting to put the milk back in the fridge. Elle thinks her father's failing health might have been more noticeable to her dad than to the rest of the family. His own father had gone through the same thing when he had been around Elle's age. Pretty soon, her father got the news: he had Huntington's disease, and his two adult daughters were potentially carriers of the gene.

A year after Elle met John, she still hadn't been tested for the gene. For her, the years following her father's diagnosis, when she knew she had a fifty-fifty chance of having the gene, were the toughest. By the time she and John decided she should be tested for the sake of their future offspring, she was resigned to the idea that she had it. A positive result would simply be a confirmation. A negative result would be winning the lottery. Elle's test results came back positive. Her twenty-nine-year-old sister hasn't been tested yet. She might be if she ever decides to have children, but for now, she doesn't want her life impeded by the weight of knowing.

Elle wasn't sure she wanted to have children until John asked her to. He told her he didn't want to be left alone. I can imagine both the tenderness and the terror of such a conversation. Elle came to the realization that no matter who you are or what you are contending with, having a child is an act of hope. Life becomes awfully bleak in the absence of hope. She agreed to try. Henry was born in March 2016.

Elle has said to me over dinner that knowing she has Huntington's feels surreal. She keeps wondering what she should be doing differently. "I guess," she finally said, "it's enough to just like your life."

I agree. It has to be enough.

Sometimes, it comes to me as a tightening in my chest. Other times, it is a rapid expansion with the simultaneous sensation of being flooded, like a dam has opened up over a giant hole where water and air have come to merge.

I will never reproduce.

It doesn't arrive as an emotion . . . it is neither happy nor sad. It is a physical truth. It exists in my cells—in fact, *it is cellular*. From a one-celled organism to the thirty-seven trillion cells of the human body . . . reproduction is *their* yearning, not ours. Whether or not my quiet mind wants children, wants motherhood, wants my heart to exist outside of my body, is not a choice.

Desire is sometimes a quiet crusader, unable even to name the thing that sits between our lives and our cellular quest for reproduction. To say we desire to reproduce is sometimes as good as our language can get. Mostly, I would say that I do not want to reproduce;

it's just that sometimes the urge comes to me, unbidden. It awakens in my cells, where it lives all the time—until I contract into oblivion, or expand beyond the boundaries of our universe—and I flood.

I will never reproduce.

It is neither sad nor happy. It is a physical truth. It lives in my cells. I have learned to let the physical truth overtake me and then I get on with my life.

THIRTY-SIX

A recent newspaper article written by a well-intentioned reporter referred to me as a person who "carries a genetic mutation in her heart that will eventually cause her a long and painful death."

As a writer, I love the sentence. "A genetic mutation in her heart" is simply stunning in its poetry. That this heart mutation will cause a long and painful death, well . . . it's grim even if it's possible. I mean, death is death, but to have one that is long and painful . . . that's another story. As the person referred to in the article, however, I have a different set of feelings.

The good news here is that the article is wrong, and I'm not just saying this because who the hell wants a sentence about them to predict a long and painful demise? The truth is that like everyone else, I have as much of a chance of getting hit by a car and dying swiftly as I do of developing a cancer that has nothing to do with my variant gene—in my heart or otherwise. Even if I live out my life without being stricken by any other malady or having an accident, I have as much chance of *not* having a long and painful death as I do of having one. It's true my father's death was the opposite of short and painless, but my ninety-one-year-old grandmother rarely had pain in her life.

Then there is the simple truth that there is no such thing as a "mutation in a heart," mine or anyone else's. Hearts don't mutate.

Genes do.

My family's mutation, or *variant*, which is a nicer word, is too new to have given rise to a body of knowledge about it. Medicine is and has always been based on study and observation over time. We just haven't had enough time or study to observe or understand how this variant works.

We have found a way to "cure" the world of our scourge. For the first time in human medical history, we don't have to wait for evolution to run with our variant, or quietly select it into oblivion. We can use the mind-blowing powers of science. Seven people have died of this gene. Seven people continue to live with it. So far, no one in the sixth generation has been passed my family's gene.

Our work isn't done. Mistakes can still be made. There are four people in my generation with the gene, and all of them can still have children. My cousin Denise doesn't plan to have any more, but if she were to accidentally become pregnant, her Jehovah's Witness faith would preclude abortion. If Hilary or I became pregnant, we might decide that our drive to give birth trumps our need to weed out the gene. We have an even younger cousin with the gene who has not yet found a partner, but who may or may not be interested in IVF with preimplantation genetic selection. In other words, life happens. Nothing is certain until it is.

But it is a plan, and so far, the plan is holding steady.

As for us, the ones who carry the family gene, how might we continue to survive? Right now we all take a pill every day that helps to keep our abdominal blood pressures low, reducing the risk of a serious bleed. Valerie was never on these pills, which might have kept her from suffering a stroke. We are all taking the allergy pill that contains a chemical that Dr. Kricket found had some impact on people with asthma, an illness with a different variant on the same gene. Perhaps it will keep our good cysteinyl-leukotriene receptor keys in our livers matching up with those cysteinyl-leukotriene locks for longer.

Meanwhile, at my last doctor's visit after my stroke-turned-post-panic-attack, I learned that my latest MRI showed that my spleen is shrinking. Dr. Sigal, that preeminent liver specialist, looked at me with bewilderment and said, "I've only ever seen this in people who have undergone successful liver transplants."

Dr. Kricket continues to maintain that Hilary and I, as women, have great potential to live long and healthy lives.

"It is still true," she insists, that "men have one X chromosome, women have two. You have one that is normal, one that is abnormal. This is why I've always told you and why I still believe women have the disease less severely."

I didn't know then that the answer was that my pressures are improving. Maybe it's the asthma pill. Maybe it's the blood pressure meds. Maybe it's genetic—but things, for now at least, are looking up.

―――――――――

My grandmother Shirley died in December 2015. She was ninety-one years old. The causes of her death included lymphedema, but as my aunt Kathy pointed out, she was ninety-one and her heart stopped. My grandma spent her whole life telling Hilary and me that we would live to be old ladies like her. She never had a blockage in her liver, but who knows? Maybe she'll be right.

Meanwhile, the Seidman lab remains poised to publish their findings. Once they do, a much larger dialogue with the medical and medical genetics communities can take place. We are hopeful that by sharing this information, we will not only find help ourselves, but we will also help others with rare vascular and lymphatic conditions.

Our bodies are wondrous. Perhaps that's all we ever need to know.

REFERENCES

American Association of Clinical Chemistry Website. Lab Tests Online, "Albumin." Last modified April 8, 2016: https://labtestsonline.org /understanding/analytes/albumin/tab/test/.

AmericanPregnancy.Org Official Website. "First Trimester Screen." Last updated July 2015: http://americanpregnancy.org/prenatal-testing /first-trimester-screen/.

Amsel, Sheri. "Where Is Your DNA?" *Exploring Nature Educational Resource: Life Science, Earth Science, and Physical Science Resource for K-12*, accessed March 2015: http://www.exploringnature.org/db /detail.php?dbID=106&detID=2456.

Anand, Geeta. *The Cure: How a Father Raised $100 Million and Bucked the Medical Establishment in a Quest to Save His Children.* New York: William Morrow Paperbacks, 2009.

Annas, George, J.D., M.P.H., and Sherman Elias, M.D. *Genomic Messages: How the Evolving Science of Genetics Affects Our Health, Families, and Future.* New York: HarperOne, 2015.

Arney, Kat. "Tracking Down the BRCA Genes (Parts 1 and 2)." *Cancer Research UK*, February 28–29, 2012.

Assaf, Zoe. "Stanford at the Tech: Understanding Genetics, Ask a Geneticist." *Tech*, September 21, 2012.

Baker, Billy. "Research, Marriage Link Singular Duo." *Boston Globe,* September 3, 2007.

Bateson, William, and Gregor Mendel. *Mendel's Principles of Heredity.* London: Cambridge University Press, 1909.

Bock, Christoph. "Why Are Women Stripy." I Fucking Love Science via Wikimedia Commons: http://www.iflscience.com/plants-and -animals/why-are-women-stripy.

BreastCancer.Org Official Website. "Mastectomy Risks." Last updated May 15, 2013: http://www.breastcancer.org/treatment/surgery /mastectomy/risks.

CancerResearchUK.org Official Website. "How Cancer Starts." Last updated October 27, 2014: http://www.cancerresearchuk.org/about -cancer/what-is-cancer/how-cancer-starts.

Chemical Heritage Foundation Official Website. "James Watson, Francis Crick, Maurice Wilkins and Rosalind Franklin": http://www .chemheritage.org/discover/online-resources/chemistry-in-history /themes/biomolecules/dna/watson-crick-wilkins-franklin.aspx.

Collins, Francis S., M.D., Ph.D. National Human Genome Research Institute. "World Economic Forum: A Brief Primer on Genetic Testing." Last updated April 30, 2013: https://www.genome .gov/10506784/.

Cooley, Denton A., M.D. "C. Walton Lillehei, the 'Father of Open Heart Surgery.'" *Science Volunteer,* September 28, 1999: http://circ.aha journals.org/content/100/13/1364.full.

Crewdson, John. "Gene Test That Went Awry Was Concealed." *Chicago Tribune,* March 9, 1997.

Dawkins, Richard. *The Extended Phenotype: The Long Reach of the Gene.* London: Oxford University Press, 1999.

———. *The Selfish Gene.* London: Oxford University Press, 2006.

Deford, Frank. *Alex: The Life of a Child.* New York: Viking Adult, 1983.

DeVita, Vincent T. Jr., M.D. and Elizabeth DeVita-Raeburn. *The Death of Cancer: After Fifty Years on the Front Lines of Medicine, a Pioneering Oncologist Reveals Why the War on Cancer Is Winnable—and How We Can Get There.* New York: Sarah Crichton Books, 2015.

Dobrzynski, Judith H. "The Lost Years of Woody Guthrie: The Singer's Life in Greystone Hospital." *Aljazeera America,* January 12, 2014.

Dominus, Susan. "The Mixed-Up Brothers of Bogotá." *New York Times Magazine,* July 9, 2016.

Donaldson, Susan James. "Fugates of Kentucky: Skin Bluer Than Lake Louise." *ABC News Official Website*, February 22, 2012: http://abcnews.go.com/Health/blue-skinned-people-kentucky-reveal-todays-genetic-lesson/story?id=15759819.

Doring, Gerd, and Niels Hoiby. "Early Intervention and Prevention of Lung Disease in Cystic Fibrosis: a European Consensus." *Journal of Cystic Fibrosis*, 2004: https://www.elsevier.com/__data/promis_misc/2004.pdf.

Facing Our Risk of Cancer Empowered Official Website. "No One Should Face Hereditary Cancer Alone." Last updated November 5, 2015: http://www.facingourrisk.org/understanding-brca-and-hboc/information/risk-management/oophorectomy/basics/overview.php.

Farber, Seymour M., and Henry R. Eagle. "Streptomycin Therapy of Tuberculosis." *Western Journal of Medicine*, July 1948: http://www.ncbi.nlm.nih.gov/pmc/articles/PMC1643317/?page=1.

Fife-Yeomans, Janet, and Ian Walker. "How a School Conversation Exposed the Incest of the 'Colt' Family." *Daily Telegraph*, June 12, 2014.

Gifford, Bill. *Spring Chicken: Stay Young Forever (or Die Trying)*. New York: Grand Central Publishing, 2015.

Gioia, Ted. *Love Songs: The Hidden History*. London: Oxford University Press, 2015.

GoFundMe. "Saving Eliza." Last updated May 20, 2016: https://www.gofundme.com/ElizaONeill.

Granados, José A. Tapia. "Life and Death During the Great Depression." *Proceedings of the National Academy of Sciences*, October 13, 2009.

GuestHollow.com Offical Website. "Extracting DNA from Strawberries Experiment." November 7, 2012: http://guesthollow.com/blog/2012/11/extracting-dna-from-strawberries-experiment/.

Haldeman-Englert, Chad, M.D., FACMG. "Autosomal Dominant." *U.S. National Library of Medicine*. Last updated May 5, 2015: https://medlineplus.gov/ency/article/002049.htm.

———. "Sanfilippo Syndrome." *U.S. National Library of Medicine*. Last updated April 20, 2015: https://medlineplus.gov/ency/article/001210.htm.

Harvard University Official Website. "BBS Faculty Member—Christine Seidman." Last updated August 22, 2013: http://www.hms.harvard.edu/dms/bbs/fac/SeidmanChristine.php.

Hathaway, Millicent L. "Trends in Heights and Weights." *Yearbook of Agriculture*, 1959.

Hirschler, Ben. Reuters. "TIMELINE—Milestones in gene therapy." April 27, 2015: http://www.reuters.com/article/health-genetherapy -timeline-idUSL5N0XK41J20150427.

Howard Hughes Medical Institute Official Website. "Christine E. Seidman, M.D., Investigator, 1994–Present." Last updated May 2, 2016: http://www.hhmi.org/research/genetic-causes-human-heart -disease.

Hughes, Mark. Interviewed by Norah O'Donnell. "Breeding Out Disease." *60 Minutes, ABC*. October 26, 2014.

Hurt, Raymond. "Tuberculosis Sanatorium Regimen in the 1940s: A Patient's Personal Diary." *Journal of the Royal Society of Medicine:* http://www.ncbi.nlm.nih.gov/pmc/articles/PMC1079536/.

Johnson, Mark, and Kathleen Gallagher. *One in a Billion: The Story of Nic Volker and the Dawn of Genomic Medicine.* New York: Simon & Schuster, 2016.

Keating, Caitlyn. "Family of Girl with Rare Disorder Quarantine Themselves for Nine Months." *People*, February 19, 2015.

Keim, Brandon. "Gene Therapy Cures Color-Blind Monkeys." *Wired*, September 16, 2009.

———. "Gene Therapy Restores Sight." *Wired*, September 22, 2008.

Kroll, David. "Angelina's Round Two with Mutated BRCA1: Solid Science Sprinkled with Nonsense." *Forbes*, March 26, 2015.

Lewis, Ricki. "Founder Populations Fuel Gene Discovery." *Scientist*, April 16, 2001.

Lorenzo's Oil, directed by George Miller. Universal City, CA: Universal Studios and Kennedy Miller Productions, 1992.

Mayo Clinic Official Website. "Huntington's Disease." Last updated July 24, 2014: http://www.mayoclinic.org/diseases-conditions/huntingtons -disease/basics/symptoms/con-20030685.

Medscape Website. "Cavernous Sinus Syndromes." Last modified February 14, 2014: http://emedicine.medscape.com/article/1161710 -overview.

Morse, Stephen P. "From DNA to Genetic Genealogy: Everything You Wanted to Know But Were Afraid to Ask." First printed in the *Association of Professional Genealogists Quarterly*, March 2009: http://www.stevemorse.org/genetealogy/dna.htm.

National Human Genome Research Institute. "Genetic Mapping." Last
    updated October 21, 2015: http://www.genome.gov/10000715.
Nelson, Bryce. "Blue People–Appalachian Marvel." *La Crosse Tribune,*
    November 6, 1974.
Nova. "Cut to the Heart." April 8, 1997: http://www.pbs.org/wgbh/nova
    /heart/.
————. "Secret of Photo 51." April 22, 2003: http://www.pbs.org/wgbh
    /nova/photo51/.
Nova Official Website. "Pioneers of Heart Surgery." Last modified April 8,
    1997: http://www.pbs.org/wgbh/nova/body/pioneers-heart-surgery.html.
————. "Rosalind Franklin's Legacy." Last modified April 22, 2003:
    http://www.pbs.org/wgbh/nova/tech/rosalind-franklin-legacy.html.
Parrington, John. *The Deeper Genome: Why There Is More to the Human
    Genome Than Meets the Eye.* London: Oxford University Press, 2015.
Pitt, Angelina Jolie. "Angelina Jolie Pitt: Diary of a Surgery." *New York
    Times,* March 21, 2015.
————. "My Medical Choice." *New York Times,* May 14, 2013.
Radiolab. "Antibodies, Part I: CRISPR." Produced by Molly Webster.
    Director of Sound Design, Dylan Keefe. Written by Jad Abumrad and
    Robert Krulwich. National Public Radio, June 6, 2015.
Ridley, Matt. *Genome: The Autobiography of a Species in Twenty-Three
    Chapters.* London: Fourth Estate, 1999.
————. *Genome: The Autobiography of a Species in Twenty-Three Chapters.*
    New York: Harper Perennial, 2006.
Sexton, Buck. "Mystery Solved: 'Blue Faced' Appalachian Family Caused
    by Rare Medical Condition." *Blaze,* February 16, 2012.
Shapiro, Beth. *How to Clone a Mammoth: The Science of De-Extinction.*
    Princeton: Princeton University Press, 2015.
Shears, Richard. "In the Valley of the Damned: Over Four Generations,
    One Extended Family Practiced a Cult of Incest Hidden from the
    World." *DailyMail.com* Official Website, December 13, 2013:
    http://www.dailymail.co.uk/news/article-2523555/Horror-Australian
    -incest-cult-spanned-generations-revealed.html.
Sifferlin, Alexandra. "Lessons from Storm Sandy: When Hospital
    Generators Fail." *Time,* October 30, 2012.
Simmons, Rebecca, Ph.D. "Epigenetic Influences and Disease." *Scitable by
    Nature Education,* 2008: http://www.nature.com/scitable/topicpage
    /epigenetic-influences-and-disease-895.

Solomon, Andrew. *Far from the Tree: Parents, Children, and the Search for Identity*. New York: Scribner, 2013.

Stolberg, Sheryl Gay. "Gene Therapy Ordered Halted at University." *New York Times*, November 28, 1999.

———. "The Biotech Death of Jesse Gelsinger." *New York Times*, January 22, 2000.

Sullivan, Brian. "Day 24: Elizabeth Sprague Coolidge." *Berkshire Eagle*, January 23, 2011.

Swartzendruber, Donna, M.S.N., R.N., C.N.N. "What Is Albumin?" *DaVita Official Website*, August 2, 2014.

Trost, Cathy. "The Blue People of Troublesome Creek: The Story of an Appalachian Malady, an Inquisitive Doctor, and a Paradoxical Cure." *Science 82*, November 1982.

University of Arizona Official Website. "Charles L. Witte, M.D. (1935–2003): Quintessential Lymphologist and Reluctant Surgeon." *Witte Tribute, Lymphology 36* (2003): http://www.u.arizona.edu/~witte/contents/wittetribute.pdf.

University of Pittsburgh Medical Center Official Website. "Low Platelet Count (Thrombocytopenia)." Accessed March 2015: http://www.upmc.com/patients-visitors/education/cancer/pages/low-platelet-count.aspx.

U.S. Department of Health and Human Services Official Website. National Institutes of Health. "All About the Human Genome Project (HGP)." Last updated October 1, 2015: https://www.genome.gov/10001772/.

———. National Institutes of Health. "Genetic Testing: How Is It Used for Healthcare": https://report.nih.gov/nihfactsheets/ViewFactSheet.aspx?csid=43.

———. National Institutes of Health. "National Cancer Institute: BRCA1 and BRCA2 Cancer Risk and Genetic Testing." Last updated April 1, 2015: http://www.cancer.gov/about-cancer/causes-prevention/genetics/brca-fact-sheet.

———. National Institutes of Health. "National Cancer Institute: Surgery to Reduce the Risk of Cancer." Last updated August 12, 2013: http://www.cancer.gov/types/breast/risk-reducing-surgery-fact-sheet.

———. National Institutes of Health. "What Is Asthma?" National Heart, Lung and Blood Institute. Last modified August 14, 2014: http://www.nhlbi.nih.gov/health/health-topics/topics/asthma.

————. National Institutes of Health. "Thoracentesis." National Heart,
Lung and Blood Institute. Last modified July 30, 2014:
http://www.nhlbi.nih.gov/health/health-topics/topics/thor.

U.S. National Library of Medicine. National Institutes of Health.
"Transjugular Intrahepatic Portosystemic Shunt (TIPS)." Last updated
February 11, 2015: https://medlineplus.gov/ency/article/007210.htm.

————National Institutes of Health. "What Are the Types of Genetic
Tests?" Last modified July 19, 2016: https://ghr.nlm.nih.gov/primer
/testing/uses.

Wade, Nicholas. "Patient Dies During a Trial of Therapy Using Genes."
New York Times, September 29, 1999.

WebMD Official Website. "Banti's Syndrome." Last updated May 28,
2015: http://www.webmd.com/hepatitis/bantis-syndrome.

YourGenome.Org Official Website. "Of Mice and Men." Last updated
June 13, 2016: http://www.yourgenome.org/stories/of-mice-and-men.

Yudell, Michael, Dorothy Roberts, Rob DeSalle, and Sarah Tishkoff.
"Taking Race Out of Human Genetics." *Science,* February 5, 2016:
http://science.sciencemag.org/content/351/6273/564.

Zielenski, J. "Genotype and Phenotype in Cystic Fibrosis." *Pubmed.gov*
(2000): http://www.ncbi.nlm.nih.gov/pubmed/10773783.

ACKNOWLEDGMENTS

Ten years ago, my friend, writer Caroline Palmer told me that a story I'd told her about my dad would make a good book. For that and all the reading and discussing she's done on and about this project ever since, I am grateful. In the capable hands of the outstanding agent and amazing human Mollie Glick, it finally became a reality. I also want to thank its early champions, Sean Gannet, Molly Lyons, and Joelle Delbourgo.

Endless thanks goes to Hilary Redmond for seeing something in my story. To Denise Oswald for stepping in and captaining a boat she'd never seen before with all the skill and ability of a lifelong pilot. To everyone at Ecco, including Emma Janaskie, Dan Halpern, Beth Parker, Miriam Parker, Ashley Garland, and Meghan Deans, for their invaluable input.

I have to thank the many brilliant writers with whom I share a writing group who have basically been this book's first, second, and eightieth editors including Sam Ritchie, Mary Atkins, Christine Clarke, Kate Tellers, Nicole Solomon, Joanne Solomon, TJ Wells, Jorge Novoa, Jessica Mannion, Sivan Butler-Rotholz, Amit Wehle,

Alia Phibes, Ilise Carter, Alina Simone, Philip Willcox, Aaron Wolfe, and Elizabeth Tannen. Also thank you to its early readers, Ann Fischer, Ellen Doyle, Christine O'Malley, Erica Barth, Allison Deutermann, Sarah Willcox, Marge Bender, Keith Newland, Jacob White, Jen Spitzer, with special thanks going out to Susan Fuhrman, Abby Sher, Boomer Pinches, and Susannah Cahalan.

An enormous and raucous *THANK YOU* goes to Dr. Peter Sabatini for translating all of that science. To Nadine Thorson for bringing me to Peter, and to Susan Faulkner for bringing me to Nadine. To Molly Anderson, R.N., Dr. Jennifer Robbins, and Michelle Gilats for your time, edits, and brilliance. To Dr. Keith Pattison, Dr. Steven Delaveris, Dr. Marlys Witte, Dr. Larry Lynn, Kim Kneuvin, Terry Silverman, Dr. Michael Landzberg, Margaret Vild, Dr. Anita Steinberg, Dominique Sartain, Evelyn Garbowit, Rabbi Harold Berman, Rabbi Harold Salzmann, Audrey Salzmann, Lucas Elijovich, and Ryan Miller for your memories of my family and your important impact on our lives. To Bo Bigelow, and Glenn and Cara O'Neill— thank you for all you do for rare genetic diseases. Thank you to Zack O'Malley Greenburg, Frank Deford, Alisa Kermisch, Elijah Saintonge, Joy Larsen Haidle, Todd Toler, Jen Difiglia, Suzanne Arnold, Nina Frenkel, Libby Batten, Sharon Cohen, Jill Barnett, Connie and Darrell Luikart, and Kerry Luikart. Thank you to Dr. Samuel Sigal, Dr. Hearns Charles, Dr. Ziv Haskel, and Dr. Mitchell Spinnell for all you do for my family and your contribution to our lives and stories.

To all the members of my family who relived many sad memories in order to bring our family and illness to life in this book, especially cousins Val and Mike, Sue and Michael, Phyllis and Mark, Kenny and Joanne, Suzanne and Vinnie, Denise, Richie and Dolly, Meryl, Jeff, Rachel, Marcus, Lizzie, Danny, Abby, Jordan, Rebekah, Rachel, and to my aunts, Joanie, Enid, Kathy, and Ellen.

My love and thanks go to my dear friends Amy Huck, Jason Evege, James Babson, Lisa Switkin, Jackie and Jimmy Shulman, and